THE OXFORDSHIRE YEOMANRY

The front and back cover are taken from a watercolour of 1896 by Orlando Norie.
The colour of the uniform facings is crimson, but soon after this the
'Mantua Purple' of which the Regiment was so proud was adopted.

The Story of

THE OXFORDSHIRE YEOMANRY

Queen's Own Oxfordshire Hussars

1798 – 1998

David Eddershaw

OXFORDSHIRE YEOMANRY TRUST

1998

First published in 1998 by The Oxfordshire Yeomanry Trust, TA Centre,
Oxford Road, Banbury OX16 9AN.

0 9534694 0 9

Printed by Oxuniprint
Oxford University Press

Oxfordshire Yeomanry Trust

This account of the County Yeomanry's History was commissioned by the Oxfordshire Yeomanry Trust.

The Trust is a registered charity, Number 1068207, with three purposes:

1. To generate and foster public interest in the Yeomanry—also known as the Queen's Own Oxfordshire Hussars—by assembling, preserving and making available for public display, the Trust's Regimental Collection.
2. To stimulate a strong sense of pride and regimental affiliation in serving members of the Regiment and to encourage the recruitment of volunteers.
3. To assist past and present members of the Queen's Own Oxfordshire Hussars and their dependants who may be in need as a result of illness, old age or misfortune

The Trustees are either past members of the Queen's Own Oxfordshire Hussars or individuls with a keen interest in the Regiment. The Trustees also include, ex officio, Her Majesty's Lord Lieutenant for Oxfordshire and the Officer Commanding the present reserve unit incorporating the title 'Queen's Own Oxfordshire Hussars'.

The Trust is anxious to receive donations or loans of items for its collection and will make good use of financial gifts. Those interested in supporting the Trust by gift or loan of artefacts, or by gift of money, are invited to contact the Honorary Secretary at the TA centre, Oxford Road, Banbury OX16 9AN.

CONTENTS

Acknowledgements

The Oxford Preservation Trust for its generosity in financing the preparation of this book, which has been produced in connection with the bi-centenary exhibition at the Oxfordshire County Museum, Woodstock.

The author would like to thank the following for assistance in locating the records of the regiment and other background material:

Staff at the Oxfordshire County Record Office.

Dr Malcolm Graham and staff at the Centre for Oxfordshire Studies.

Martin Allitt and staff at Banbury Reference Library.

Staff at the Chipping Norton branch of the Oxfordshire Library Service.

Staff at the National Army Museum Reading Room.

Hugo Brunner, Lord Lieutenant of Oxfordshire and other members of the Oxfordshire Yeomanry Trust, especially Col. Tim May and Maj. Bob Sheldon, who have collected items from other sources and have personally provided information on the recent history of the regiment.

"Nick" Nicholls, former member of the QOOH who has freely made available his extensive collection of information on the regiment's history, including his own reminiscences and others collected from Ben Pritchett, Maj. Barnett and Col. John Thomson.

The National Army Museum for permission to reproduce photographs from their collection.

Oxfordshire County Record Office for permission to reproduce documents from their collection.

The President and Fellows of Magdalen College, Oxford, for permission to reproduce the painting of the Prince Regent's Visit.

John Murray (Publisher) Ltd for permission to quote from "The Oxfordshire Yeomanry in the Great War" by Adrian Keith-Falconer.

Joan Brasnett for preparing photographs for publication.

Paul Cleal for typesetting and layout.

Main Sources

Records of the Lord Lieutenancy held at Oxfordshire CRO, catalogue references L/M.I to L/M.X.

Register Books and Regimental Order Books of the Oxfordshire Yeomanry and Queen's Own Oxfordshire Hussars held by the Trust.
Keith-Falconer, A. "The Oxfordshire Hussars in the Great War" (Murray, 1927).
Peel, the Hon.S. "Trooper 8008 IY" (Arnold, 1901).
Nicholls, T.E. "Yeomanry Memories" (Privately printed, 1992).
Oxfordshire newspapers held at the Centre for Oxfordshire Studies.
The Victoria County History of Oxfordshire, vol x, Banbury.
Photographs and printed ephemera held at the National Army Museum.
Scrapbooks kept by Viscount Valentia and Col.John Thomson.
Reckitt, B.N. "The Lindley Affair" (privately printed 1972).

Bibliography of some general works relating to the Yeomanry and their military, political and social background.

Beckett, I. F. W. "The Amateur Military Tradition, 1558–1945" (Manchester, 1991).
Charnley, J. "Churchill, The End of Glory" (Hodder, 1993).
Churchill, W. S. "My Early Life" (Collins, 1930).
Eddershaw, D. G. T. "Chipping Norton—The Riot of 1845" (In TopOxon vol ix, 1973/4).
Emery, F. "Marching Over Africa: Letters From Victorian Soldiers" (Hodder 1986).
Farwell, B. "For Queen and Country: A Social History of the Victorian and Edwardian Army" (Lane 1981).
Gann, L. H. and Duigan, P. "The Rulers of British Africa 1870–1914" (Croom-Helm, 1978) (Chapters on the army and military organisation).
Herbert, G. "Shoemaker's Window".
Macdonald, L. "1914, The Days of Hope" (Penguin).
Mileham, P. "The Yeomanry Regiments, 200 Years of Tradition" (Canongate Academic, 1994).
Peters, C. "The Lord Lieutenants and High Sheriffs of Oxfordshire" (Perpetua Press, 1995).
Reaney, B. "The Class Struggle in Nineteenth Century Oxfordshire, A Social Background to the Otmoor Disturbances" (History Workshop, 1970).
Trinder, B. "Victorian Banbury".

I The Origins of The Oxfordshire Yeomanry

War with France: Home Defence and the Militia

In the eighteenth century Britain possessed only a small standing army and no professional police force. Both were traditionally viewed with great suspicion as being expensive and a threat to the constitutional freedom of British citizens. The idea of full conscription, which was introduced in some continental countries and produced large national armies, was so unpopular in Britain that no government dared suggest it. (It was not introduced until 1916). There was also the feeling that Britain's real defence was the Navy and the seas which formed a natural barrier to invasion. Yet the need for a home defence force not only to protect the islands from foreign invasion, but also from domestic riot and tumult, had long been recognised. The British army consisted of a comparatively small number of regular regiments for use anywhere in the world, plus the county militia and various volunteer forces raised to meet particular emergencies for home defence. Several invasion scares resulting from wars with France and Spain, together with the Jacobite Rebellions in 1715 and 1745, led to the re-organisation of the militia in 1757. The government ordered the Lord Lieutenant of each county to raise a specified number of men for the militia each year. In theory every able-bodied male between 18 and 50 was liable to serve and the selection was made by means of a ballot, but there

Muster Roll of the Persons enrolled of of Oxfordshire Yeomanry Cavalry
George Earl of Macc-

Effective and Exempt from the Militia &c — Enrolle

1	Earl of Macclesfield	Shirburn	1798
2	Willm Lowndes Stone Esq	Brightwell	
3	Richard Clerke Esq	Kingston Blount	
4	Henry Newberry	Bensington	
5	Willm Coyens	Watlington	
6	John Fane Esq	Stokenchurch	
7	Edward Good	Aston Rowant	
8	John Hayward	Watlington	
9	Thos Greenwood	Turville Bucks	
10	Joshua Stratton	Watlington	
11	John Hamp	Do	
12	Willm Hayward	Do	
13	William Scoles	Shirburn	
14	William Wiggins	Watlington	
15	James Burgess	Bensington	
16	Henry Ansell	Swincombe	
17	Charles Greenwood	Easington	
18	William Pitkin	Stokenchurch	
19	Aeden Stevens	Chalford	
20	Jonathan Wellar	Brightwell	
21	Richard Swell	Newington	

were several classes of exemption and the better off could pay a substitute if they were unlucky enough to be chosen, or might even be prepared to pay the £10 fine for refusal to serve. Militiamen did 28 days training a year and were equipped and paid for this at public expense. In time of national emergency they were required to serve in any part of the UK. While some men volunteered or went as substitutes for the sake of the pay and adventure, many poorer families who could not afford to buy exemption lived in fear of losing a loved one and bread winner, and government measures to increase the numbers of the militia because of the Napoleonic wars led to widespread rioting in many parts of Britain.

The militia had none of the social cachet of later yeomanry regiments and was generally despised by the upper levels of county society. Although only those with land worth at least £50 per annum could become officers, it was inevitable that the lower ranks came to be filled mainly by men of the poorer classes, and when it came to quelling popular outbursts against bread prices or other hardships they were not always thought to be reliable. Men of the Oxfordshire Militia themselves rioted against low pay, the high price of flour and the need to use it for dressing their pigtails, when camped near Brighton in June 1795. After breaking open local stores and forcing bakers to sell at reasonable prices, about 200 men were arrested and the ringleaders condemned to death. At Goldstone Bottom on the 14th of June Private Henry Parish of Chipping Norton and Private Edward Cooke of Witney were executed in front of the assembled regiment by a firing squad drawn from their fellow rioters.

1. **The earliest surviving roll of the Watlington Troop. This troop was raised in 1798 and was the first to be raised in Oxfordshire. The list shows the names of some of the original members, including George, Earl of Macclesfield. Most of the men were drawn from the area around the Earl's seat at Shirburn Castle.** *(Oxfordshire Archives)*

Volunteers, Fencibles and Yeomanry

Faced with the desperate need for more troops to undertake home defence in the 1790s when there seemed a very real threat of Britain being invaded by the armies of revolutionary France, it was decided to increase the size of the militia and to raise additional volunteer forces. Those in authority, both at national and local level, feared not only invasion from France, but infiltration of revolutionary ideas among disaffected members of the British working classes, such as those discussed at the meetings of the London Corresponding Society and in other radical groups. The Habeus Corpus Act was suspended making the detention of suspects easier and large political meetings were banned. It was expected that an important role of the new volunteer forces would be the suppression of riots and tumults and they were therefore intended to be recruited from respectable property owners and gentlemen whose loyalty could be relied upon. The Government made an appeal for the raising and financing of volunteer regiments in 1794. A meeting of "Nobility, Gentry, Freeholders and Yeomanry of the County" was called at the Star Inn in Oxford and in the same year a regiment of Fencible Cavalry (a temporary force for the duration of the current emergency) was raised by the Hon. Thomas Parker. In a number of other counties regiments of Yeomanry were raised.

The war against France went badly at first: the British were driven out of the Low Countries and the Coalition of Britain's allies fell apart. Now the fear of invasion increased and further pressure was applied to the counties to raise troops, especially cavalry. There was already a tax on horses, but the Provisional Cavalry Act of 1796

2. **Illustration in a Regimental Album recording the foundation of the original troops of Yeomanry in Oxfordshire in 1798. It was drawn nearly a century later and mistakenly shows the officers of the Wootton and Watlington Troops combined into one troop.** (*Courtesy of the National Army Museum*)

1798.

PRIVATE of YEOMANRY CAVALRY *"Left Parry!"*

"OXFORDSHIRE REGIMENT"
of
"Gentlemen & Yeomanry Cavalry"

— 1798. —

Captains	Oldfield Bowles	6 Sep 1798
	G. Et of Macclesfield	20 June 1798.
Lieutenants	W. Sam Churchill	29 Nov 1798.
	W. Lowndes Stone	20 June 1798.
Cornets	J. Duffell.	29 Nov. 1798.
	R. Clerk	20 June 1798.

Agent. Learmonth & Buzley. Parliament St.

BANBURY & BLOXHAM.
"TROOP"
of
"Gentlemen & Yeomanry Cavalry"

— 1798 —

Captain. Thos Cobb ..... 27 Sep 1798.
Lieutenant J. Davies ..... 14 Jan. 1799.
Cornet. G. Councer ..... 14 Jan. 1799.
Surgeon J. Shorter ..... 14 Jan. 1799.
Agent. Mr Hemans. Whitehall.

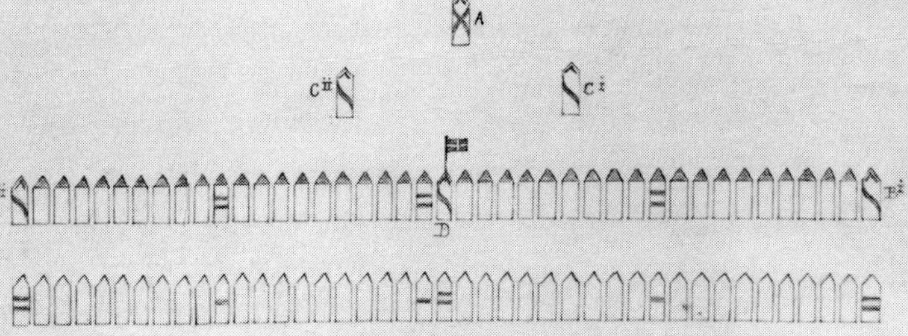

THE SQUADRON 36 FILES.

— 1794 —

A { Offr Comg SQUADRON. 2½ Horses Length fr Squadron.
BI Senr Offr Comg Rt Qr Rank
BII Next Senr Offr Comg Lft Qr Rank.
CI Next Senr Offr Comg 2nd Qr Rank.
CII Next Senr Offr Comg 3rd Qr Rank.
D Cornet with Standard.
Sergeant ═ Corporal ─

made a further requirement for the listing of all horse owners in the county and one in every ten horses was liable to be requisitioned for military service. Further meetings of the Lieutenancy took place at the Star Inn, and lists of horses and owners in each hundred and parish were drawn up. Rates of allowances for uniforms and upkeep of horses were fixed. This was an unpopular system because it appeared to treat the gentry in the same way as ordinary militiamen who were regularly posted away from their home counties, and might require the more substantial farmers to serve alongside their inferiors without due respect for social distinction

The way out was offered by the formation of volunteer units by those who wished to demonstrate their undoubted patriotism but avoid the disadvantages of the militia ballot and central government control. Such units were often formed by the prosperous middle-class element in the towns, who may also have wished to demonstrate their independence of the county landed gentry. The Oxford Loyal Volunteers were founded in 1796 and lasted until the temporary peace of 1802. The rapid growth of such units, often more concerned with the protection of property on a very local scale than with repelling invasion or reinforcing the regular army, received a mixed response from the government.

Much more welcome were the volunteer cavalry units raised by the nobility and gentry of the counties and recruited from people who shared the same patriotic and social values. "Yeomanry" was the title universally applied to such corps, who saw themselves as the successors to the medieval yeomen thought to have been the backbone of English society and English armies in the past.

The sense of urgency increased when in February 1797 a small French invasion force actually landed near Fishguard and was rounded up by an even smaller local force including a recently formed troop of yeomanry. (Tradition also claims that the French may have been mislead by the coloured capes of the Welsh women spectators into thinking that a regiment of Redcoats was approaching.)

The Oxfordshire Yeomanry

In 1798 the first four yeomanry troops were formed in Oxfordshire. The lead was taken by George, Earl of Macclesfield, who founded the Watlington Troop in May or June of that year with himself as Captain (the highest rank possible for a small unit.) William Lowndes Stone of Brightwell Park (High Sheriff of Oxfordshire in the previous year) was Lieutenant and Richard Clerk Cornet. Three further troops were soon formed and by the end of 1798 there were four quite independent units with about fifty men in each. They seem to have been based on the "Hundreds"—ancient administrative divisions of the county with a deputy lieutenant in charge of each. The new troops were the Bullingdon, Dorchester and Thame Troop (Captain Sir Christopher Willoughby), The Wooton Troop (Captain Oldfield Bowles) and the Bloxham and Banbury Troop (Captain Sir Thomas Cobb). Oldfield Bowles was a remarkable man whose interests ranged from music and fine art to progressive methods of farming—a leading figure locally in the agricultural changes of the late eighteenth century. Thomas Cobb of Neithrop came from a family whose wealth was derived from wool but who lost a good deal of it in the next generation through an unfortunate venture into banking.

These units did not form a county regiment, being entirely independent of each other, although they seem to have worn similar blue uniforms with white facings and white breeches, a large white scarf and and black leather cavalry helmet with bearskin crest and side hackle plume. They were armed with swords and pistols. Each man provided his own horse, which immediately limited the type of recruit, and they may have had to meet part of the cost of the uniforms and equipment. Some government subsidy later became available, but it was not unusual for gentry and others who raised troops to spend considerable sums of their own money in equipping them. During the periods of training and official duty they

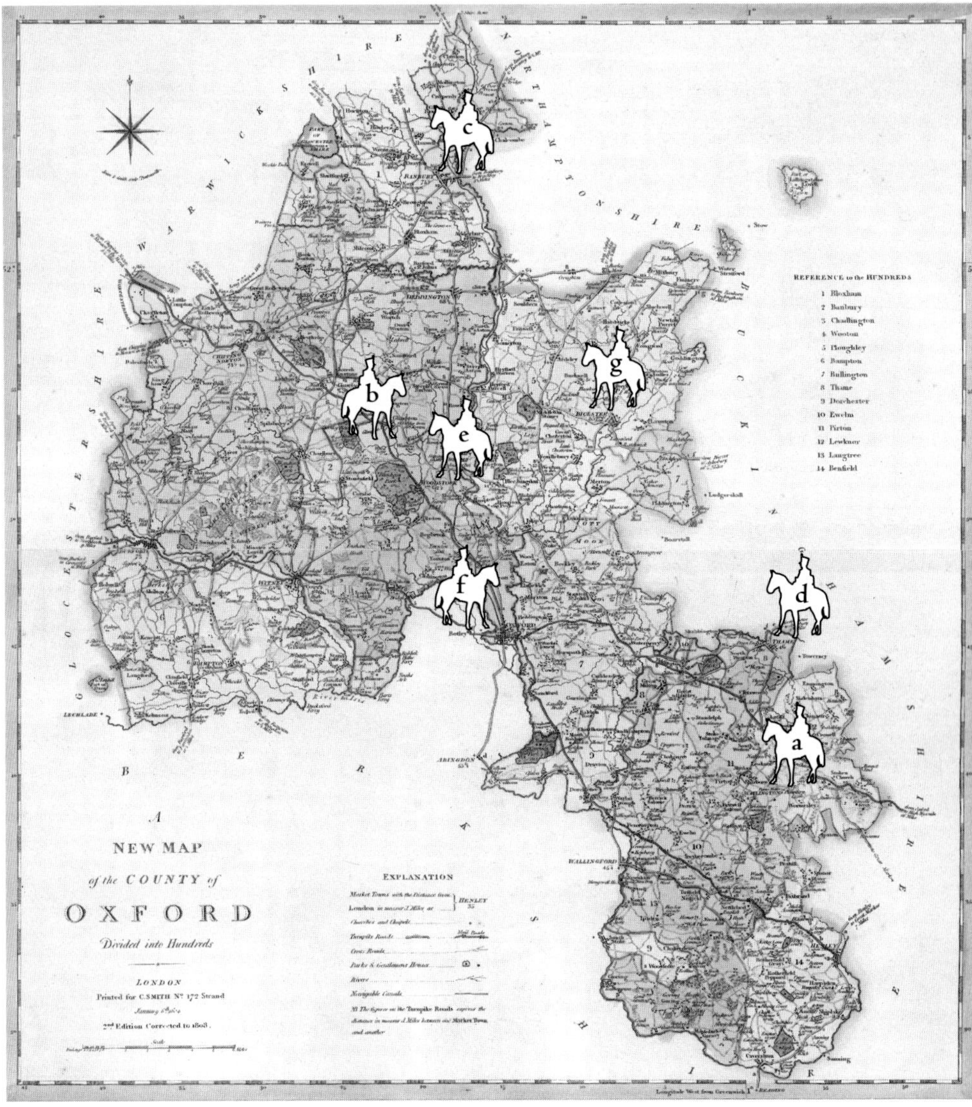

3. **The first troops of Yeomanry in Oxfordshire.** (*Graham Smith*)

a **Watlington** 1798
Capt. George, E. of Macclesfield

b **Wootton** 1798
Capt. Charles Oldfield Bowles

c **Bloxham and Banbury** 1798
Capt. Thomas Cobb

d **Bullingdon, Dorchester and Thame** 1798
Capt. Sir Christopher Willoughby, Bart.

e **Woodstock** 1803
Capt. Francis Spencer (Lord Churchill)

f **Oxford** 1803
Capt. Henry Whorwood

g **Ploughly & Bicester** 1803
Capt. John Harrison

were paid by central government. Later in the century this training amounted to eight days a year but during the war it was more frequent. However, because a large proportion of troopers were farmers, training periods tended to avoid haymaking and harvest and other busy times in the farming year. Spurred by a sense of patriotism and a desire to protect private property and the established order of society, there was also the additional advantage of being exempt from the militia ballot and from the horse tax. By 1801 there were about 21,000 yeomanry enrolled throughout Britain.

1802 appeared to bring an end to the war with France under the terms of the Peace of Amiens. Most yeomanry regiments were disbanded but some, including the four Oxfordshire troops, took the option offered by the government of continuing in existence without central funding and entirely at their own expense, although they continued to enjoy exemption from horse tax. It proved only a shortlived peace and preparations for the renewal of war began before the year was out. The government urgently needed to increase the size of the regular army and to make the militia more effective, but the idea of forming independent volunteer forces was much more popular in the country. So many were formed that the government was for a time unable to cope with the cost of arms and ammunition for their training or with the additional administration involved in enrolling them. It was more welcoming to yeomanry, who were felt to be more reliable. The Lord Lieutenant held another meeting at the Star Inn and made a plea to all loyal citizens to enrol as volunteers "in the military and yeomanry corps already established". He reminded them of "the threats of an inveterate foe to invade this land, with intent to overthrow and annihilate its constitution in church and state . . . . and his intention of giving up to rapine and plunder the persons and property of the inhabitants of this happy isle".

This was enough to inspire the formation of three more troops of yeomanry in Oxfordshire: the Oxford Troop (Captain Henry

Whorwood), the Ploughley or Bicester Troop (Captain John Harrison) and the Woodstock Troop (Captain Lord Spencer). In addition to these new foundations, an extra troop was recruited for the existing Bloxham and Banbury Troop making it into a squadron, now commanded by Major-Commandant G F Stratton, a gentleman who had made a fortune in the service of the East India Company and eventually retired to the manor of Great Tew. The establishment of all troops was increased to 80 men. The 7 units remained independent even to the extent that some developed different uniforms, the Woodstock Troop appparently wearing scarlet with yellow facings. Extravagant uniforms were to be a feature of the yeomanry tradition in the nineteenth century. Another feature of the yeomanry in Oxfordshire was the connection with the Spencer-Churchill family, beginning in 1803 and lasting well into the twentieth century.

With the exception of the Bicester Troop which disbanded in 1807, the Oxfordshire troops continued their separate existences until the end of the Napeolonic War, fulfilling their training obligations and establishing a sense of loyalty to their troop and to the yeomanry ideal. When Oldfield Bowles died in 1810 his funeral hearse was accompanied all the way from Oxford to North Aston by the Wootton Troop, "his sumpter horse, with boots and spurs, being led by two of the troop in solemn funeral procession." Perhaps the yeomanry of those days were better suited for this kind of impressive ceremonial than for actual warfare. They were fortunately not called upon to face an invasion by Napoleon's formidable Army of England.

II In Support of the Civil Power

Suppressing Riots and Tumults

From its inception the yeomanry had been saddled with the duty to turn out when required to support the civil authorities in suppressing riots, protecting property and maintaining order, a duty to which it was felt such men would be instinctively drawn by their own position as property owners in the upper ranks of county society. Governments probably saw the yeomanry as far more effective in this role than as a home defence force to resist invasion, and the number of yeomanry units increased after the war when regular establishments were being reduced. It certainly proved to be its most important role in the first half of the nineteenth century. It has also remained the most controversial aspect of yeomanry history.

The image of the "Peterloo Massacre" in Manchester in 1819 has coloured popular opinion of the yeomanry ever since. Called on by nervous local magistrates to arrest the speakers at a huge working class meeting in St Peter's Fields, mounted yeomanry and regular troops used their swords against the protesters. Eleven people were killed and many wounded in the panic. The government later congratulated the magistrates on their action.

The Bloxham and Banbury Squadron in Action

In the restless and difficult years following the end of the Napoleonic Wars, Oxfordshire like the rest of Britain experienced protests accompanied by violence and lawbreaking, usually motivated by the poverty and frustration of the poorer members of society, unrepresented in Parliament and seeming to have no other way of improving their lot. Even as early as 1800, while the war was still in progress, there had been riots in Banbury over low wages and the high price of food. The Red Lion Inn was attacked because it belonged to a prosperous farmer and corn dealer. At a subsequent public demonstration, mounted troops charged the crowd and a child was killed. When the ringleader of the rioters was escorted to Oxford further rioting broke out there, foreshadowing the Otmoor affair of thirty years later.

1830 was the worst year. Impoverished labourers turned to rick burning and machine breaking in the so-called "Swing Riots" all over southern England. Trouble again occurred around Banbury when local men protested against low wages, unemployment and the use of threshing machines by farmers, which led to more unemployment. A machine was smashed and set on fire at night by a crowd of men at Neithrop and part of the Bloxham and Banbury Squadron of Yeomanry under Major Stratton and Captain Thomas Cobb was summoned to quell the riot. The labourers, however, attacked the Yeomen with stones collected from a nearby newly repaired road, and gathered in their smocks like apples in a girl's apron. Some also used long burning poles taken from the blazing machine, which so frightened the horses that their riders were unable to control them. Major Stratton fell off and his men were

4. **Part of a page from the Register Book of the First Regiment of Oxfordshire Yeomanry Cavalry commanded by Lieut-Col. the Right Hon. Lord Francis Churchill. This entry records six days when a troop was stationed at Islip "In Aid of the Civil Power" during the Otmoor enclosure riots of the 1830s.** (*Oxfordshire Yeomanry Trust*)

chased ignominiously all the way back to Banbury. An eye-witness has left a lively account of this event which highlights the frightening ordeal faced by inexperienced volunteers even when dressed up in the impressive uniforms of the Yeomanry.

> "The soldiers were of no use as they soon had to make a hasty retreat—in fact they were completely beaten off the field. They mounted and rode off up into the town as if the Devil was after them. Some were maimed and some even lost their swords. One poor fellow, the name of Lines, he belonged to the band and used to play the serpent, but he had to fight with the rest and by some means got parted from the others. He had to run the gauntlet right through the mob from the field and was awfully pelted with stones to North Bar."

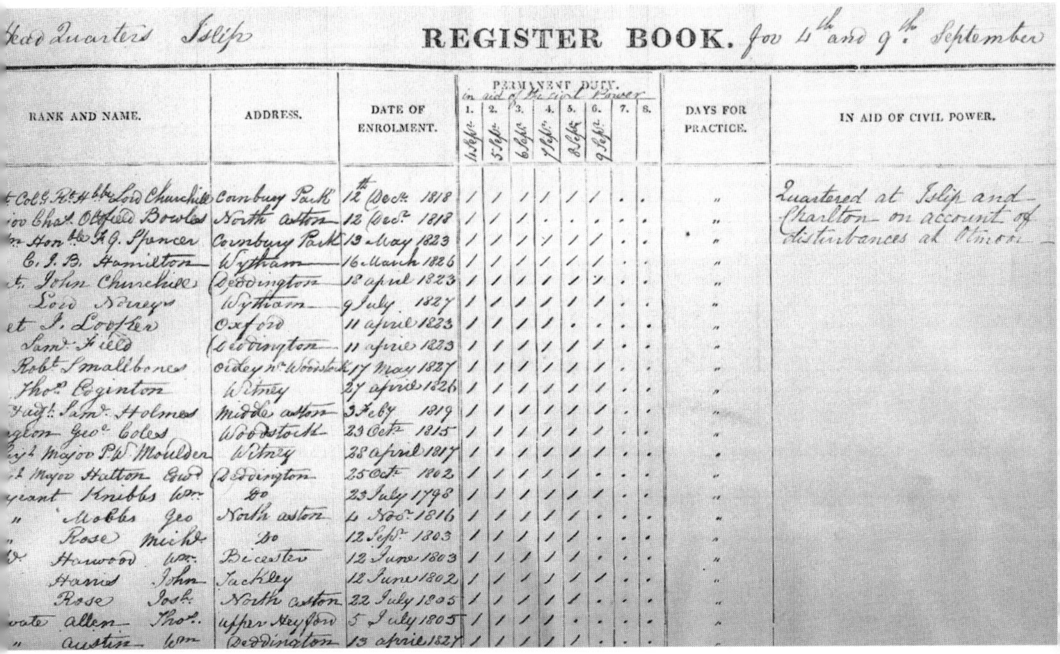
...ead Quarters Islip — REGISTER BOOK. for 4th and 9th September

RANK AND NAME.	ADDRESS.	DATE OF ENROLMENT.	PERMANENT DUTY. in aid of the Civil Power 1. 4 Septr	2. 5 Septr	3. 6 Septr	4. 7 Septr	5. 8 Septr	6. 9 Septr	7.	8.	DAYS FOR PRACTICE.	IN AID OF CIVIL POWER.
...t Col. Rt. Hble Lord Churchill	Cornbury Park	12th Decr 1818	/	/	/	/	/	/	.	.	"	Quartered at Islip and Charlton on account of disturbances at Otmoor
...or Chas. Oldfield Bowles	North Aston	12 Decr 1818	/	/	/	.	.	.	.	.	"	
...n Honble F. G. Spencer	Cornbury Park	13 May 1823	/	/	/	/	/	/	.	.	"	
C. J. B. Hamilton	Wytham	16 March 1826	/	/	/	/	/	/	.	.	"	
...t. John Churchill	Deddington	18 April 1823	/	/	/	/	/	/	.	.	"	
Lord Norreys	Wytham	9 July 1827	/	/	/	/	/	/	.	.	"	
...et J. Looker	Oxford	11 April 1823	/	/	/	/	.	.	.	.	"	
Saml. Field	Deddington	11 April 1823	/	/	/	/	/	/	.	.	"	
Robt. Smallbones	Oddey nr Woodstock	17 May 1827	/	/	/	/	/	/	.	.	"	
Thos. Eginton	Witney	27 April 1826	/	/	/	/	/	/	.	.	"	
...Adjt. Saml. Holmes	Middle Aston	3 Feby 1819	/	/	/	/	/	/	.	.	"	
...gton Geo. Coles	Woodstock	23 Octr 1815	/	/	/	/	/	/	.	.	"	
...t Major P. W. Moulder	Witney	28 April 1817	/	/	/	/	/	/	.	.	"	
...t Major Hatton Edwd	Deddington	25 Octr 1802	/	/	/	/	/	.	.	.	"	
...geant Knibbs Wm	Do	23 July 1798	/	/	/	/	/	/	.	.	"	
" Mobbs Geo	North Aston	4 Novr 1816	/	/	/	/	/	.	.	.	"	
" Rose Michl	Do	12 Sepr 1803	/	/	/	/	/	.	.	.	"	
...r Harwood Wm	Bicester	12 June 1803	/	/	/	/	/	.	.	.	"	
Hand John	Jackley	12 June 1802	/	/	/	/	/	/	.	.	"	
Rose Josh	North Aston	22 July 1805	/	/	/	/	/	.	.	.	"	
...vate Allen Thos	Upper Heyford	5 July 1805	/	/	/	/	.	.	.	.	"	
" Austin Wm	Deddington	13 April 1827	/	/	/	/	/	.	.	.	"	

The rioters were met in the town by a strong force of tradesmen and others sworn in as special constables, and these together with some regular Dragoons who later arrived from Coventry eventually stopped the trouble for that night. On the following night there was another incident in Tadmarton but this was dealt with far more effectively by the Dragoons whose horses were trained for action. When the rioters again tried to stampede them with burning poles "this was no use against the regular soldiers as they rode right up to the fire and would have leaped over it if required".

The first Reform Bill was hotly debated in Parliament in 1831, accompanied by even hotter agitation all over the country, culminating in fierce riots in places like Bristol and Nottingham, where the castle was burned down. Banbury was again the scene of much excitement and some fighting, but the radical Mayor decided not to use the Yeomanry this time, either because he thought their mere presence might provoke trouble, or for fear that they might intimidate radical voters in the council election, which was said to have happened elsewhere. Barricades were placed across the streets to keep the Yeomen out.

Otmoor for Ever!

The years 1830 and 1831 were anxious times in Oxfordshire and the courage of the Yeomanry volunteers was repeatedly put to the test. While Major Stratton's men were fighting rioters in Banbury, Lord Churchill's enlarged squadron formed from the original Wootton and Woodstock Troops, and now called the 1st Oxfordshire Regiment of Yeomanry Cavalry, was called on to deal with the even more persistent defenders of the ancient common rights of Otmoor.

Attempts by Sir Alexander Croke of Studley and others to enclose and drain Otmoor and share it between private landowners had been going on unsuccessfully since 1788. He was vigorously opposed by many of the smaller farmers and in particular by the poorer inhabitants of the seven small townships around the moor, whose ability to survive independently of parish relief or wage labour depended on their ancient right to graze animals and gather fuel on this common land shared by them all. At last, in 1815, an Act for the Enclosure of Otmoor was passed by Parliament and the land divided up between 73 owners, the largest of whom included Croke himself, Lord Abingdon, the Duke of Marlborough, three Oxford Colleges and four local Rectors. Work progressed over the next fifteen years to enclose and drain this lowlying moor.

In 1830 a number of the small farmers became angry when a new cut designed to divert the River Ray and help the moor to dry out, repeatedly overflowed and flooded their crops. Twenty-nine of them took the law into their own hands and broke down the banks of the cut intending to restore the river to its original course. Brought before the magistrates in Oxford, they were acquitted as having a reasonable grievance, and that encouraged many of the poorer inhabitants to follow their example.

In September large gangs of men began to go out onto the moor at night, often with blackened faces and sometimes disguised in women's clothes, to carry out concerted attacks on the fences and hedges. Horns were sounded, there were shouts of "Otmoor for ever!" and great cheers went up as each fence was destroyed. Croke's farm was attacked and his son knocked to the ground when he threatened to shoot at the rioters. Parish constables were powerless or unwilling to intervene. Both the militia and Lord Churchill's Troop of Yeomanry Cavalry were called out, 148 officers and men of the Yeomanry being stationed at Islip and Charlton on Otmoor. The following day, the 6th September, about 1,000 men, women and children openly marched around the Moor "Repossessioning" it, breaking down fences and appointing their own herdsman to regulate the grazing of their livestock. The Magistrates appealed for order, then read the Riot Act and ordered the Yeomanry to disperse the crowd and arrest the leaders. Forty-four men were captured and

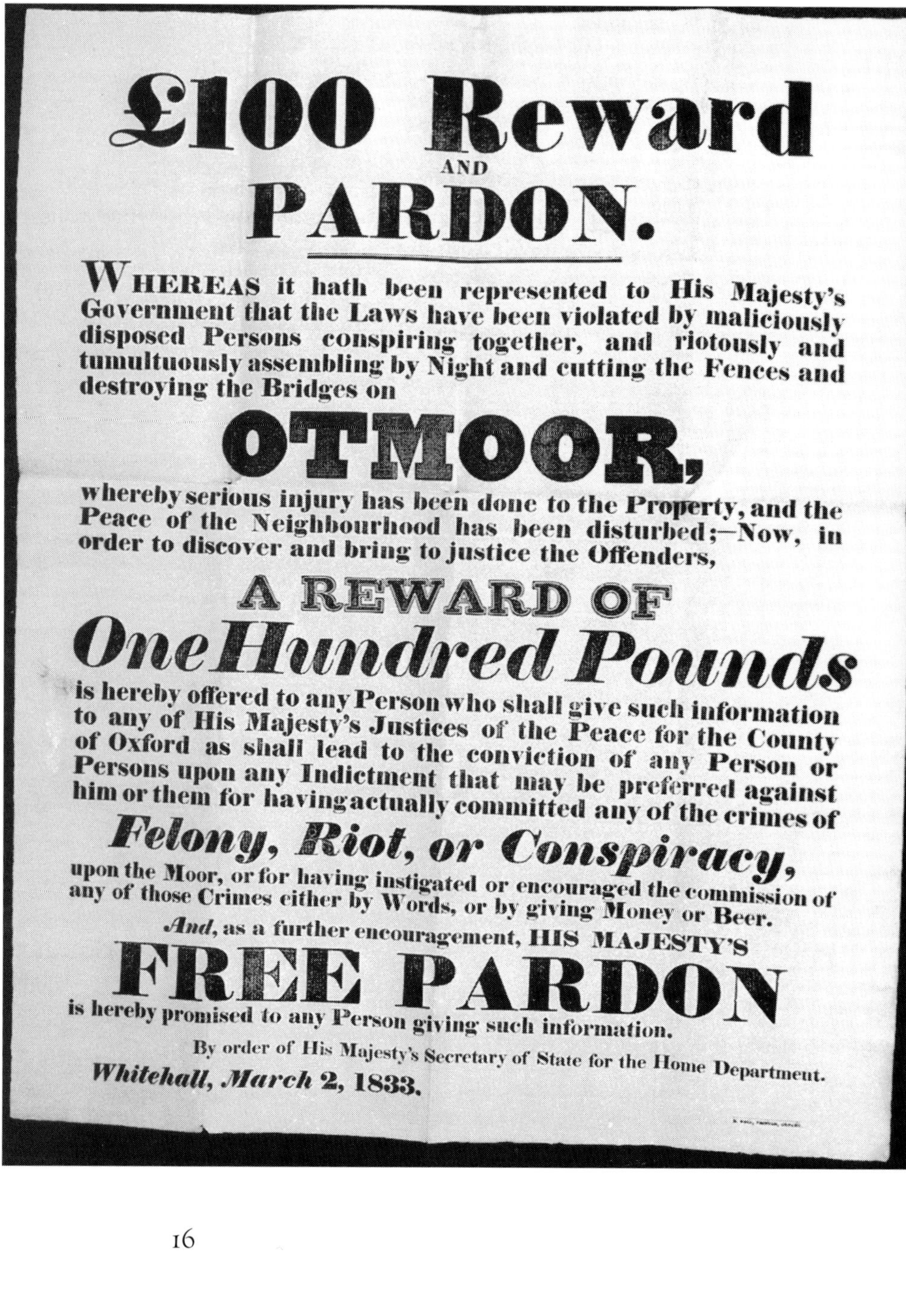

£100 Reward AND PARDON.

WHEREAS it hath been represented to His Majesty's Government that the Laws have been violated by maliciously disposed Persons conspiring together, and riotously and tumultuously assembling by Night and cutting the Fences and destroying the Bridges on

OTMOOR,

whereby serious injury has been done to the Property, and the Peace of the Neighbourhood has been disturbed;—Now, in order to discover and bring to justice the Offenders,

A REWARD OF

One Hundred Pounds

is hereby offered to any Person who shall give such information to any of His Majesty's Justices of the Peace for the County of Oxford as shall lead to the conviction of any Person or Persons upon any Indictment that may be preferred against him or them for having actually committed any of the crimes of

Felony, Riot, or Conspiracy,

upon the Moor, or for having instigated or encouraged the commission of any of those Crimes either by Words, or by giving Money or Beer.

And, as a further encouragement, HIS MAJESTY'S

FREE PARDON

is hereby promised to any Person giving such information.

By order of His Majesty's Secretary of State for the Home Department.

Whitehall, March 2, 1833.

ordered into two farm waggons to be taken under escort to Oxford gaol. The escort consisted of four militia sergeants, with a cornet, two sergeants and twenty-two privates of the Oxfordshire Yeomanry, under the command of Captain Hamilton.

As they approached Oxford, news of the riot at Otmoor having gone before them, they were met by a hostile crowd throwing stones and insults at the soldiers. Coming nearer to the city they had to pass between the stalls and sideshows of St Giles Fair, forcing a way through a crowd of several thousand working people thronging the wide street. This crowd, described by *Jackson's Oxford Journal* as "vast numbers of the worst description of people" now turned away from the fair to the greater excitement of attacking the Yeomanry and helping the unbound prisoners to escape. Beaumont Street was at that time under construction and afforded a ready supply of bricks and stones which were hurled at the vastly outnumbered Yeomen. Captain Hamilton reported later that at this point "the attack was greatly increased and rendered it impossible to keep our ground without making a decided charge upon the rioters, which from their number would only have uselessly endangered the lives of my men". One of the Yeomen tried to use the flat of his sword but had his ear split, a tooth knocked out "and was carried away from his post nearly fainting from loss of blood". The *Morning Chronicle* confirmed that "had the soldiers fired, not one of them would have escaped with his life". Meanwhile the prisoners got away into the crowd and the soldiers fled to cries of "Unhorse them! Unhorse the buggers! Down with them!" Volunteer yeomanry were simply not equipped by training or temperament to counter such concerted and orchestrated opposition.

The Otmoor troubles continued for several years and the Oxfordshire Yeomanry were called out on at least five other occa-

5. **Government poster offering large rewards to informers during the Otmoor enclosure riots.** (*Oxfordshire Archives*)

sions to deal with rioters. About 40 new men were recruited from Bicester and the area to the north of Otmoor forming a separate troop within the regiment in 1831. Clearly there was some support for their role. At one time Yeomen were on extended duty patrolling the moor in relays for up to three weeks. A troop of the Buckinghamshire Yeomanry was also used and Lord Melbourne, the Home Secretary, was persuaded to send detachments of Dragoons and Coldstream Guards at various times. Finally a special force of police brought from London was set up. None of these measures was wholly successful. They clearly demonstrate the difficulties of the Yeomanry's role at this time.

The Chipping Norton police riot, 1845

Individual Yeomen may sometimes have had doubts about this kind of service. There is perhaps a hint of this in their final experience of supporting the civil powers in Oxfordshire, which happened at Chipping Norton. In November 1845 a serious riot broke out over the action of one of the borough police, Sgt. Charles Knott. He was one of the first of the new professional police, selected from Peel's Metropolitan force, to be employed in Oxfordshire in place of the old elected parish constables and night watchmen, and he was probably already unpopular with some in the town. In 1845 he unfortunately killed a man through over-zealous use of his truncheon while arresting him during a fight in a local pub. A noisy crowd gathered in the Market Place threatening violence, and persisted for three days so that the Mayor was forced to send an urgent request for military aid to the Duke of Marlborough as Lord Lieutenant.

The Duke called out the Woodstock Troop of the Yeomanry, but to his amazement and obvious anger only Cornet Evett and one Trooper responded, followed by three more men a little later. He fired off an angry letter to the Clerk asking "what I ought to do and what I can do" with those who had failed to respond to his orders.

Meanwhile Col. Bowles from North Aston managed to raise a detachment of about 30 men and marched to Chapel House near Chipping Norton. The incident is interesting because it might suggest that some of the troopers sympathised with people in Chipping Norton who hated their new type of police sergeant. On the other hand it may simply be that it was just not possible for working farmers and others to turn out at a moment's notice. The men were

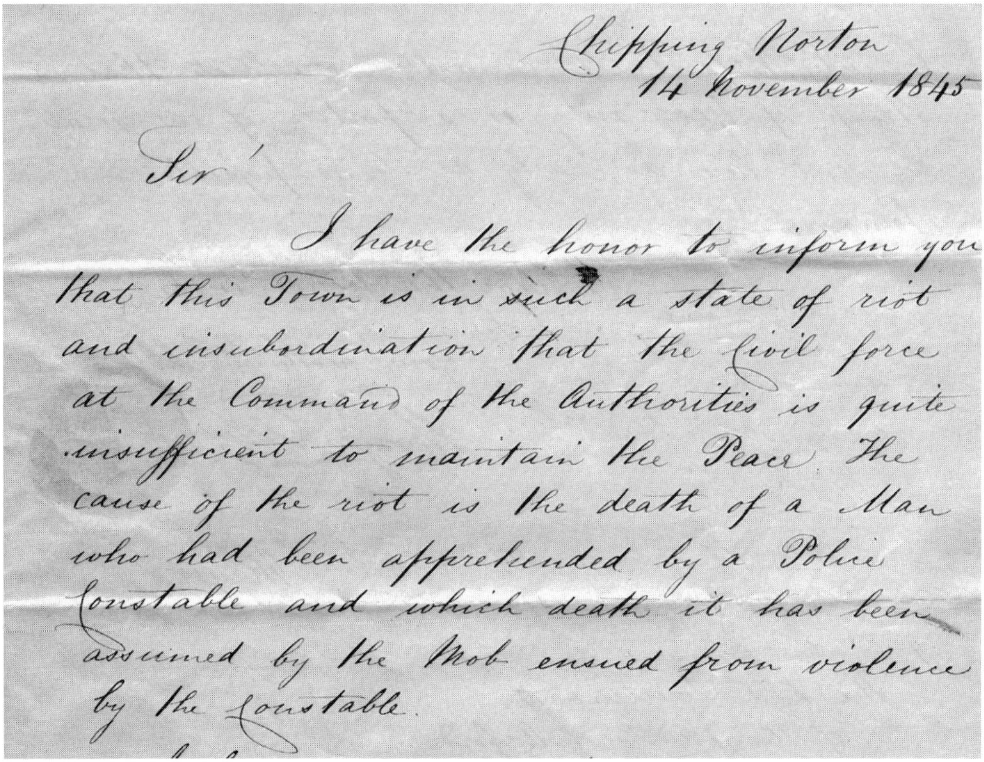

Chipping Norton
14 November 1845

Sir

I have the honor to inform you that this Town is in such a state of riot and insubordination that the Civil force at the Command of the Authorities is quite insufficient to maintain the Peace. The cause of the riot is the death of a Man who had been apprehended by a Police Constable and which death it has been assumed by the Mob ensued from violence by the Constable.

6. **Letter from the Mayor of Chipping Norton to the Lord Lieutenant of Oxfordshire requesting aid from a troop of Yeomanry because "this town is in a state of riot and insubordination" after a man had been killed accidentally by a police sergeant.** *(Oxfordshire Archives)*

questioned about their failure and their excuses can still be seen scribbled on the back of a letter among the Lord Lieutenant's papers. Some were ill, some "away from home"; several claimed they could not get a horse and some did come later. Two men clearly refused but we can only guess at their reasons. They did not earn much sympathy from Mr Morrell who wrote to the Duke expressing his opinion that "there is nothing like having the Yeomanry ready when called upon, and I would send those Woodstock devils to the rightabout!"

Meanwhile, Sgt Knott was sent to Oxford gaol to await trial on a charge of manslaughter and was escorted for his own safety by Col. Bowles's men as far as Woodstock. He was eventually acquitted.

This was the last time that the Oxfordshire Yeomanry were called on to support the civil power in suppressing riots and tumults. The role fell in future to Sgt Knott's successors in the new county constabularies set up after 1856, and as standards of living improved gradually for the working classes riots became less frequent. The experiment of using part-time yeomanry troops to deal with riots had not proved altogether successful. This was not because of any lack of courage or determination on their part, but although associated closely with the class against whom most riots were directed and therefore likely to pursue their role with vigour, their few days annual training, concentrating mainly on formal and ceremonial drills, did little to equip them for the ordeals they had to face when called out by the civil powers.

III "The Queen's Own"

Consolidation of the original troops

In 1817 Lord Churchill, brother of the 5th Duke of Marlborough and commander of the Woodstock Squadron, increased his squadron's size to three troops, and a year later the original Wootton Troop under Captain Oldfield Bowles amalgamated with the Woodstock to form "The North-West Oxfordshire Regiment of Yeomanry Cavalry". A few years later this was being referred to as "The First Oxfordshire Regiment" as it began to recruit more widely. Central government had been urging for some time that separate units should merge into single county regiments but the Oxfordshire troops jealously preserved their independence. The Watlington, Oxford and Bullingdon Troops continued their separate existence for another six years until 1824.

Meanwhile the Bloxham and Banbury Squadron was also enlarged from two to three troops in 1824. Maj Stratton wrote from Park Hall, Alcester, to the Clerk of the Lieutenancy explaining that "it must now be called the Bloxham and Banbury *Corps* of Oxfordshire Yeomanry, as being 3 troops it is now more than a squadron tho' less than a regiment".

Faced with mounting costs, and some disquiet about the role of the yeomanry in dealing with civil disturbances, the government ordered the disbanding of most regiments in 1828. Both the First

Oxfordshire and the Bloxham and Banbury resisted this by obtaining royal approval for their continued existence, on the understanding that it would be without pay or allowances and therefore at no cost to the government. Within a couple of years, however, government was again anxious to have the yeomanry available to put down rioting labourers in many parts of England and the First Oxfordshire Regiment was restored to the official list in 1830. Perhaps because of its dispiriting experience in the face of the machine breakers in Neithrop in that year, the Banbury and Bloxham was not restored and ceased to exist as a separate unit. Captain Fane reformed the Watlington Troop at this time and it appears to have remained an independent unit at least until 1835.

Royal occasions

In June 1814 Oxford had witnessed one of the most prestigious events in its long history, when the Prince Regent paid a state visit to the University and City, accompanied by the Emperor of Russia, the King of Poland and other European dignitaries. This great occasion was to mark what at that moment seemed to be the end of the long French Wars and the final defeat of Napoleon. His eventual escape from Elba could not be foreseen, and for now England and her Allies rejoiced.

The Prince Regent dismounted from his carriage at Magdalen Bridge and walked at the head of a huge procession up High Street, followed by the Allied heads of state, (although according to *Jackson's Oxford Journal* the King of Prussia was on this occasion so plainly dressed in a brown coat that at first he was not recognised). Behind them came "a prodigious number of foreign Princes and Noblemen of the first distinction, including the Prince of Orange, the Prince of Mecklenburgh, Prince Metternich, Count Leiven, etc, etc," The elderly Field Marshal Blucher, soon to save the day for Wellington at Waterloo, was also present. Escorting this glittering procession were members of the county's yeomanry troops, certainly including those from Woodstock and Bloxham and Banbury, together with some regular Dragoons. The Journal describes the scene: "Troops of horse soldiers stationed at regular intervals close to the flag pavement, kept the middle of the street free for the procession, while the foot pavement itself was filled with well-dressed spectators, and the windows, tops of houses and battlements with ladies and gentlemen waving their handkerchiefs and hats". When the two-day visit came to an end, the *Journal* commented that "Great praise is due to Col. Lord Spencer, Vice-Lieutenant of the County, and his regimental staff, Lt-Col Weyland, Majors Stratton and Holloway and Captain Henderson; and also to the Officers of the 9th Light Dragoons who with the brigade of cavalry formed an escort for the Royal Visitors during their stay here, and prevented them being incommoded by the pressure of the multitude".

Queen Adelaide's visit, 1835

Queen Adelaide, wife of William IV, came to Oxford simply to satisfy her own desire to see the famous university city. She was a well educated German princess (it was noted that both she and her sister the Duchess of Saxe Weimer appeared able to follow the Latin address spoken by the University Orator), and knew of Oxford's reputation as a seat of learning.

This time it is clear that the Yeomanry provided her escort, without any supporting regulars. The Queen having driven from Windsor, her carriage was met a few miles outside Oxford "and the whole cavalcade was escorted by two small bodies of Lord Churchill's Regiment of Oxfordshire Cavalry and swept forward at a most rapid pace. At every village and house along the road the inhabitants thronged forth to see her Majesty, and when she arrived here, all Oxford may be said to have been in the High Street. It was not quite one o'clock when her Majesty alighted at the Angel Hotel amid tumultuous cheering. The main body of the Oxfordshire Cavalry was drawn up before the hotel". No doubt they were at their smartest, anxious to impress not only the Queen but the Iron Duke himself who was also present in his capacity as Chancellor of the University. At various times during the next two days troops of Yeomanry provided royal escorts, including a visit to Blenheim Palace on the last morning and then for the Queen's departure from the city that afternoon. Their efforts were rewarded by perhaps the greatest honour of their regimental history. Queen Adelaide instructed Lord Howe to thank Lord Churchill and "the Officers and Gentlemen under your command", saying that "The Queen will ever entertain a grateful recollection of the loyal attachment to her Majesty's person shown by this fine Corps". In his Regimental Order of the Day for October 27th 1835, Lord Churchill informed the regiment "that her Majesty has been graciously pleased to approve that in future the First Regiment of Oxfordshire Yeomanry shall bear the appellation of 'The Queen's Own' ". An announcement was duly placed in the London Gazette to this effect. Captain Fane's independent Watlington Troop of Yeomanry was also thanked, but received no additional honour.

Who were the Yeomen? Officers and Gentlemen

Yeomanry regiments are inevitably linked in popular imagination to the gentry, and indeed to the nobility, who frequently took the lead in raising troops and themselves filled the most senior positions. The Earl of Macclesfield and Lord Churchill were not exceptional in this: in neighbouring counties Lord Chandos (Buckinghamshire) and the Duke of Beaufort (Gloucestershire) filled the same role. Such noblemen took a very personal interest in their regiments, which were often referred to not by their official title but by the name of the Commanding Officer. The First Oxfordshire was usu-

7. **Officers of the Queen's Own Oxfordshire Yeomanry Cavalry, 1872. The age of some of these officers, their air of aristocratic superiority and the rather unmilitary bearing of the group, is typical of the Yeomanry at this date. The young Viscount Valentia (3rd from right) eventually became Commanding Officer and helped to improve the smartness of the regiment in the 1880s and '90s, while preserving its social prestige.** (*Courtesy of the National Army Museum*)

ally referred to in the press as "Lord Churchill's Regiment" and individual troops as "Captain Bowles's" or "Major Stratton's".

Even someone as important as a Duke of Marlborough felt his status was enhanced by command of his own regiment of cavalry. The granting of commissions and promotions in the Yeomanry also greatly increased the scope of his patronage which he could extend to other county families, or withhold, as he chose. Before Cardwell's reforms of the 1870s (and to some extent even after them) the question of military competence was barely considered when selecting officers. It was considered much more important that a man should come from the right sort of background and be a "gentleman" in the accepted sense of the word. In this the Yeomanry only reflected the attitudes current throughout the Victorian army where professionalism was positively discouraged. Any family or individual who aspired to a place in the higher ranks of county society in Oxfordshire was likely to seek a Yeomanry commission as a mark of status, and would be dependent for this upon the favour of the Commanding Officer and ultimately the Lord Lieutenant—posts which the 6th Duke combined.

In 1852 Lord Saye and Sele wrote a suitably respectful letter to the Duke to request a commission for his son:

> "Lord Saye and Sele presents his kindly respects to his Grace the Duke of Marlborough. He would be so much obliged to the Duke of Marlborough as Lord Lieutenant of the county of Oxford if his Grace would have the goodness to confer the appointment of Deputy Lieutenant of the County on his eldest son, the Hon Mr Fiennes, and if it would not be too much to ask, a Cornetry in the Oxfordshire Yeomanry Cavalry. His son has just left Oxford, and if he might be permitted to say so much, is in character and conduct everyway worthy of this mark of favour at His Grace's hands, and which would be so deservedly appreciated by son and father."

There was less formality when Lord Churchill wrote to the Duke:

> "My Dear Duke,
>
> May I recommend to you young Cotterell Dormer for a Lieutenancy vice Montgomery, who being never able to attend,

has begged me to convey to you his resignation of his commission.

Dormer is anxious to go out with us this year and will be an acquisition."

Captain Oldfield Bowles of North Aston had raised the original Wootton Troop in 1798 which later amalgamated with the Woodstock. His son succeeded him in the new regiment, starting as a Cornet in 1838 and rising rapidly to be Lt-Colonel by 1843, second in command only to Lord Churchill. When Lord Churchill died two years later, Bowles clearly expected to succeed to the full command, and his feelings on reading in *Jackson's Journal* that the Duke (who had no previous service) intended to appoint himself Colonel of the regiment, are obvious even from the restrained letter he wrote to the county clerk seeking confirmation of the report. He resigned in 1847 when the Duke, as Lord Lieutenant, was able to compensate him with an appointment as Colonel in command of the county Militia, but such a position would not have carried quite the same status as that in the Yeomanry.

The names of those holding commissions in the Oxfordshire Yeomanry in the nineteenth century read like a roll of the nobility and gentry of the county. The same names appear as Deputy Lieutenants and County Justices. Lord Valentia, Albert Brassey and Sir Robert Hermon-Hodge even combined being Members of Parliament with service in the Yeomanry (perhaps winning influential support thereby). A few professional men, solicitors, etc., are occasionally found among them, and the regiment included a surgeon and a chaplain. Ordinary regiments might be content with a local vicar but for the Oxfordshire Yeomanry no less a figure than the Very Rev. F. Paget, Dean of Christ Church, acted as Chaplain for several years at the end of the century, even continuing when he became Bishop of Oxford. His chief duty was to conduct a church parade during the annual training camp.

Other Ranks

The non-commisioned officers and troopers came from a different strata of society, and it is from them that the term "yeomanry" actually derives, rather than their aristocratic superior officers. A yeoman was originally a substantial farmer, denoting someone slightly beneath the rank of gentleman. In those of the surviving muster rolls which show occupations, farmers are invariably the largest single group, accounting for about a third of the men. They were part of the landed interest, sharing many of the attitudes of the gentry. The rest include a wide range of tradesmen and others, some no doubt directly dependent upon their local squires, like gamekeepers and estate workers for whom "volunteering" for the Yeomanry may have been virtually a condition of employment.

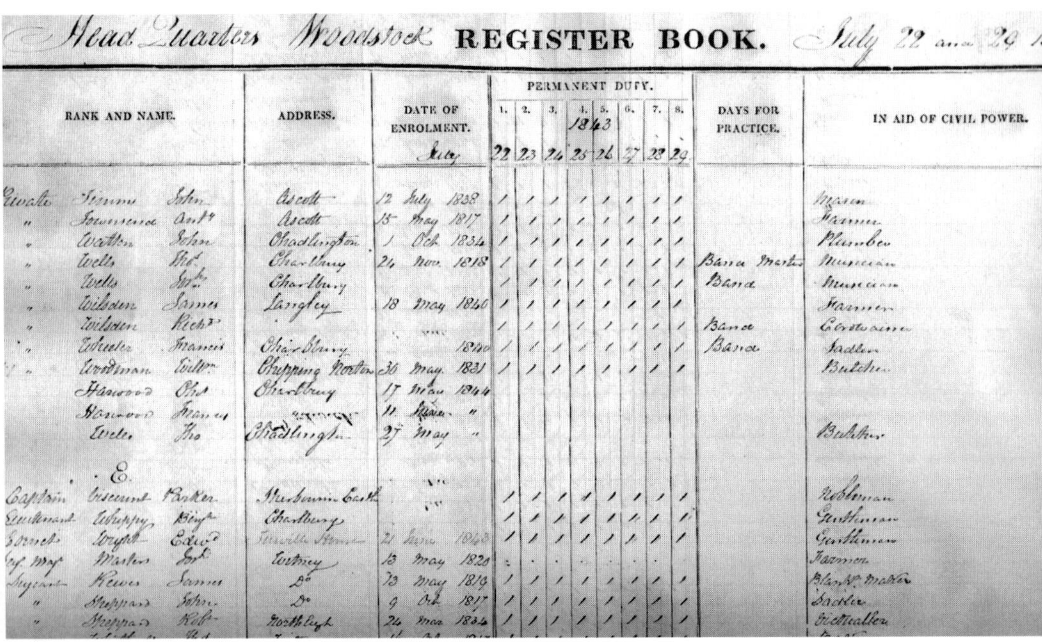

Others perhaps hoped to benefit through earning the favour of such a prestigious and wealthy group of officers. Recruiting for each troop within the regiment was often on a local basis, centred on the estate of the troop's senior officer. There were a number of farriers, smiths, sadlers and grooms who would naturally be attracted to a cavalry regiment, and many licensed victuallers. Every man had to provide a horse, but some troopers managed to borrow or hire their mounts just long enough for the annual eight days training. (The excuse "couldn't get a horse" when Yeomen failed to assemble during the Chipping Norton riot in 1845 may have stemmed from this problem). It was only in 1877 that the War Office Regulations stipulated that every man must ride his own horse "or one borrowed from some relative, friend or employer, who shall engage that the horse will be provided for the Yeoman's use whenever required."

Some men were servants from the households of the gentry and simply exchanged their livery for a Yeomanry uniform while continuing to act in this capacity as batmen to the officers. There were very few unskilled labourers; such men were more likely to join the regular infantry regiments or the militia.

Apart from any financial advantage they hoped for, many were undoubtedly motivated by what was usually called "patriotism"—a wish to uphold and defend the established order of society whose standards and outlook they shared. The spectacular uniform of the Yeomanry and the feeling of pride in belonging to such a prestigious body, gave them additional status. It was probably as much a tradition for some quite humble families, as it was for the Churchills or Fiennes, for each generation to serve in the Yeomanry. When

8. **Addresses and occupations of members of the Queen's Own Oxfordshire Yeomanry in 1843. Three members of the band come from Charlbury, two from the same family. The officers shown are a nobleman and two gentlemen, there are three farmers and several tradesmen, including a blanket maker from Witney.** (*Oxfordshire Yeomanry Trust*)

William Draper, a farrier from Charlbury, died in 1840 his son took his place in the regiment, and a few individuals served for exceptionally long periods. John Harris of Tackley served for 40 years until he died during the annual training camp in 1842, but surely the longest continual service must belong to William Knibbs of Deddington, a sadler, who enrolled when the first troops were formed in 1798 and had achieved the rank of Troop Sergeant Major by the time of his death in 1853 at the age of 75. Several other members of the Knibbs family were also long-serving Yeomen.

The essentially amateur ranks of the yeomanry were stiffened by the inclusion of a small number of regular soldiers. Each troop had a regular sergeant-major, paid by the War Office on a full-time basis, who acted as Quartermaster. They were not paid very much, however, nor could they have had much to occupy their time outside the annual training camps, so it was common practice for them to take other work. The 1877 Regulations allowed this but laid down that "under no circumstances will they be allowed to be employed in or keep public houses or licensed victuallers shops".

The band contributed much to the regiment's morale. Lord Villiers is said to have raised and trained his own band for his particular troop and this may have been the foundation of the very well turned-out band of the Oxfordshire Yeomanry later in the century. They were uniformly mounted on grey horses and included a drummer with two kettle drums and embroidered drum banners bearing the regimental insignia. The cost of such provision may have been met by the Commanding Officer personally or was shared between all the officers. Few things did more than the appearance and the sound of a good band to fill the men with pride in their regiment and to improve its public image. The Trumpeter attached to each troop had important duties beyond his part in the band, as trumpet calls signalled many orders to the rest of the troop, both on ceremonial parades and in battle. The bandsmen themselves were probably musicians in their village bands. In 1840 six bandsmen all came from

Charlbury, and there seems to have been a family tradition here as in 1843 the Wells family from this village provided the Bandmaster and two other bandsmen. Coombe provided two bandsmen from the Busby family, while a third also served though not in the band.

"Permanent Duty"

The high point of the Yeoman's year in the nineteenth century was attendance at the annual training camp, officially referred to as Permanent Duty. There were also occasional troop drills. The Permanent Duty was usually held in May or early June and timed so as not to clash with the hay harvest or other farming commitments which would be inconvenient to the large number of farmers in the regiment. It lasted for six days with a further two days allowed for travel to and from the place of the camp. For this they received pay at set daily rates ranging in 1845 from £1 12s 10d for a Colonel to 7s. for NCOs and Privates. There was also an allowance for forage for the officers' horses, and as the twenty-three officers present at the camp that year brought fifty-eight horses between them (four each for Lt-Colonels and Majors, three for Captains and Lieutenants and two for Cornets) this was an expensive item. (The Duke of Marlborough once made an inflated claim for expenses to the War Office because he mistakenly counted the number of the officers' horses instead of the officers themselves.)

In the second half of the nineteenth century it became the custom to stage the Permanent Duty at Blenheim, where the tents were pitched in the shadow of the column of victory, the expanse of the park provided space for troop exercises and the huge courtyard in front of the palace provided a suitably imposing setting for church parades and the annual inspection by a senior officer from the War Office. In spite of the imposing setting, however, the inspecting officer was not always impressed with what he saw. In 1867 he commented that "The saddlery might have more care bestowed upon it"

9. **Major Mason leads his troop in line past the Prince of Wales and Viscount Valentia, Port Meadow Oxford, 1898.** *(David Mason esq.)*

also that "the sword exercise was very indifferently performed". Particular criticisms had been made by Col. McMurdo, Inspector General of Auxilliary Forces in 1862 concerning the troopers' horses: "The men worked as well as could be expected considering that they were very indifferently mounted, the horses being very small and underbred and in some instances quite unfit for yeomanry cavalry. Considering that Woodstock is at the heart of a hunting country the men ought to provide themselves with a better class of horse."

Lord Churchill replied that it was his "anxious and unceasing endeavour to obtain a larger class of horse. But I would beg to remark that although this is a hunting country it is not a horse breeding country; that the Privates are not usually of the hunting class, and that the horses in general use are small."

The other favourite venue was Oxford, where the officers were billetted in the Randolph Hotel, and a ceremonial guard in full dress uniform paraded in front of the hotel each evening to the delight of onlookers. The Yeomen paraded each morning in St Giles and went off to Port Meadow for their training. Large crowds of locals accompanied them on their return ride down Walton Street. As well as the church parade and annual inspection, there

10. **Crowds of young men and boys accompany the QOOH along Walton Street Oxford. The mounted band is leading the regiment back to its quarters after training exercises on Port Meadow.** (*Courtesy of the National Army Museum*)

were regimental competitions and sports which also attracted crowds of spectators. The whole of the Permanent Duty became a popular public event when no doubt Yeomen of all ranks enjoyed the opportunity to strut about importantly in their fine uniforms, impressing wives and girlfriends, as well as taking part in novel forms of military training and enjoying the company of fellow volunteer soldiers. The newly popular camera has left a good record of all this from the 1890s (sadly only in black and white), especially of the glittering array of officers in their full-dress uniforms, dark blue with "mantua purple" facings and encrusted with silver braided "frogging" in ornate designs right down to the knees.

The Churchill Connection

The Spencer-Churchill family has had a long and active association with the Oxfordshire Yeomanry, commencing in 1803 with the raising of the Woodstock Troop by Lord Francis Spencer (who later became the first Lord Churchill), and continuing through several generations into the twentieth century. It is an association which has undoubtedly brought considerable benefit to the regiment. It was certainly Lord Churchill and his brother the fifth Duke who persuaded Queen Adelaide to grant the "Queen's Own" title in 1835 and there must have been many other occasions when the influence of the Marlboroughs brought advantages.

The sixth Duke appointed himself Lt-Col Commandant of the Yeomanry when Lord Churchill died in 1847. At the same time there were in the regiment the second Lord Churchill (Lt-Colonel), The Hon H Spencer (Lieutenant) and the Marquess of Blandford (Lieutenant). Lord Churchill succeeded to the command in 1857 and this involvement, often by several members of the family, continued in an unbroken tradition for more than a century. The ninth Duke was particularly active and appears in many photographs from the 1890s onwards. (There is also a much earlier photograph of him, aged three or four, in a child-sized copy of the uniform.) He first joined in the humble rank of Cornet and eventually became Lieutenant Colonel commanding the regiment from 1910 to 1914. Before that he had gone to South Africa with volunteers from the regiment and had served in a senior position on the staff of the Imperial Yeomanry during the Boer War.

Perhaps the most famous of the yeomanry Churchills was Sir Winston, apart from anything else an enthusiast in his early life for the amateur tradition and a great individualist himself. His influence led to the regiment being among the first territorials to be involved in the 1914–18 war.

Decline and recovery

During the 1870s the Oxfordshire Yeomanry suffered a short period of decline which may have been partly due to a lack of enthusiasm for Cardwell's attempts to reform and modernise the army, (although these had only a limited application to volunteer yeomanry regiments). Certainly there were many resignations—over 150 between 1875 and 1877—and in 1871 the Duke of Marlborough had written to the War Office requesting that the regiment be excused from Permanent Duty that year because, although the officers were willing, the number of men was reduced to about two thirds of the normal establishment "as a consequence of the new regulations". He went on to comment that "there is not the same *esprit* with the men as there used to be".

However, in about 1881 the rather cumbersome title of The Queen's Own Regiment of Oxfordshire Yeomanry Cavalry was changed to The Queen's Own Oxfordshire Hussars, by which the regiment has been known ever since, and this perhaps marked the beginning of a general recovery in its fortunes and reputation.

The general intention of Cardwell's work at the War Office was to improve the professional standards of the army and to increase the efficiency of its organisation. He was battling against an institution steeped in tradition and resistant to change, and perhaps no branch was more independent and resentful of War Office control than the Yeomanry. In the long term, however, they too benefited from the new measures and in particular from the influence of officers with experience in regular cavalry regiments who were used to train the Yeomanry. In 1874 a School of Instruction for officers of the Yeomanry was established at Aldershot and two years later the Certificates of Proficiency awarded there became necessary for new promotions. These involved having to pass examinations in squad drill, sword and carbine exercises and riding. From the late 1870s onwards the Oxfordshire Yeomanry included a number of former regular officers. There were several Adjutants and a succession of COs after 1878 who undoubtedly raised the efficiency of the regiment and restored its morale. They included such names as Edmund Ruck-Keene (The Bays), John Baskerville and Albert Brassey (14th Hussars), Viscount Valentia (10th Hussars) and H. C. Norris (8th Hussars). It is generally accepted that none did more to restore the standing of the regiment than Viscount Valentia who served in an active or honorary capacity for nearly sixty years, but had his greatest influence as Commanding Officer from 1894 to 1904. The improved popular image of the QOOH by the end of the century was expressed in some rather curious verses in their honour which appeared in the *Oxford Journal* in 1895:

> Go see them in the morning
> by bold Valentia led;
> Let rebels take for warning
> their manly, martial tread.
> Go see their sabres flashing
> on wide Port Meadow free.
> From Oxford's sons so dashing
> the foe would quickly flee!

Royal Connections: The Prince of Wales as Commander in Chief

In addition to his work within the regiment Valentia had very influential links outside it, holding a position as Comptroller within the royal household. Like the Duke of Marlborough he was very close to the Prince of Wales and it was clearly due to their influence, together with the high esteem in which the regiment was now held,

11. A proud group. H.R.H. The Prince of Wales (Edward VII) as Colonel in Chief, with officers of the Queen's Own Oxfordshire Hussars, 1898.

that the Prince agreed to become its Colonel in Chief in 1896, a position he retained when he became King Edward VII,—a rare distinction for a Yeomanry regiment. He inspected the regiment during its Permanent Duty camp at Blenheim in 1896 and it naturally provided his escort when he returned to Oxford in the following year to open the new Town Hall in honour of Queen Victoria's Diamond Jubilee. In the huge Jubilee procession in London, attended by crown princes, heads of state and military contingents from most countries in the world, along a six-mile route lined by 20,000 troops, those in attendance on the Prince of Wales included three representatives of the Auxillary Forces of Britain, with Viscount Valentia of the Queens Own Oxfordshire Hussars chosen to represent the Yeomanry.

IV The Boer War and Beyond, 1899–1913

The Imperial Yeomanry

Victorian expansionist policies in the second half of the nineteenth century had brought Britain into conflict on several occasions with native African states. Although her soldiers might sometimes be surprised by the fierceness of the natives and even admire their bravery, belief in their own innate superiority made eventual victory seem inevitable to the British. When the Boers, tough Dutch settlers determined to preserve their territories and their independence, resisted Britain's attempt to annex the Transvaal and gain control of its recently discovered gold mines, the government started the second Boer War with the utmost confidence that victory would be achieved quickly. This was in spite of its failure in the first war, and resounding defeat by the Boers at Majuba Hill in 1881. The attitude of supreme self-confidence was quickly shattered, however, by the succession of defeats suffered in "black week" in December 1899 when the Boers carried the attack to the British possessions of Cape Colony and Natal, and began the famous sieges of Mafeking, Kimberley and Ladysmith.

It was against this background that the War Office turned to the militia and other auxiliary forces to support the regulars. Legal obstacles to overseas service were swept aside (if only for one year, as it was expected that the war would be over in that time). The War Office's entrenched reluctance to take seriously part-time, albeit enthusiastic, volunteers like the Yeomanry was eventually overcome when General Buller called for a mounted infantry force to counter the quick moving Boers. The Imperial Yeomanry was established at the end of December 1899, with the organisation in the hands of a committee headed by three prominent commanding officers of Yeomanry regiments—Lt Col Lucas (Loyal Suffolk Hussars), Lord Chesham (Royal Bucks Hussars) and the Queen's Own Oxfordshire Hussars commander Lord Valentia.

There was great enthusiasm in the country for volunteering to go to South Africa and many times the original estimate of Yeomanry required actually went. A high proportion of these were not in fact serving Yeomen but new recruits who came forward because of the war. Horses were provided for those who did not bring their own and for those who did, there was a payment of £40. Daily pay for the men was to be 1s 3d. Over 10,000 men went in 1900 and a further 16,000 in 1901.

It was never envisaged that complete regiments of Yeomanry would be able to serve as a unit; instead those who volunteered from a regiment were formed into companies in larger units. In Oxfordshire Lord Valentia asked for 120 volunteers, a company, but those who responded formed two companies in separate battalions (the 40th Company of the 10th Battalion and the 59th Company of the 15th Battalion). This suggests that twice the number he asked for actually came forward. He was prepared to accept former members of the regiment as well as serving yeomen, "also any suitable man not an ex-member". The QOOH register for the May 1900 training camp lists 218 men on roll, of whom 47 (5 officers, 9 NCOs and 33 privates) were absent "serving with the Imperial Yeomanry", amounting to about 21 per cent. In 1901 33 were still serving (or were new volunteers) and 13 in 1902, by which time the war was nearly

The Prince Regent Received by the University and City of Oxford, 14th June 1814. This oil painting by George Jones is the earliest known illustration of the Oxfordshire Yeomanry. Members of Lord Churchill's "First Regiment" are shown on the left, controlling the crowds in High Street. *(Courtesy of The President and Fellows of Magdalen College Oxford)*

Open-air service for men of the Imperial Yeomanry near Lindley, South Africa, Sunday June 2nd, 1900.

"The next day was Sunday. We had a church parade on the slope of the hill, and we had a sermon, of which I did not hear a word, and of course we sang 'Onward Christian Soldiers'. All the time there was the occasional sound of rifle firing from the outposts."

This event during the Boer War, vividly described by Trooper the Hon. Sidney Peel (see p. 45) followed the disaster at Lindley. The service was apparently conducted by a corporal of the Imperial Yeomanry who was a curate in civilian life.

(Oil painting by Caton Woodville in the possession of the Oxfordshire Yeomanry Trust. Photography arranged by Phillips, Auctioneers, Oxford)

The culmination of a 200-hundred year tradition: a Corporal in the traditional uniform of the Yeomanry watches as the scroll granting the regiment the Freedom of Banbury is signed by the Town Clerk in the presence of the Lord Lieutenant of Oxfordshire, The Mayor of Banbury, The Honorary Colonel and members of 5 Sqdn. (QOOH), July 1998. (*Martyn Brown*)

12. **British troops parade in Church Square, Pretoria, during the Boer War. The heavy gun in the foreground is being hauled by an impressive team of 20 oxen.** (*From Visc. Valentia's scrapbook*)

over. It seems likely then, that most of those who answered Lord Valentia's call were either ex-members of the Yeomanry or new volunteers, and this matches the picture elsewhere in the country. They were motivated by the chance of excitement and adventure in a foreign country. Many young men chose this option as a short term break (for a year at the most) from routine if respectable civilian jobs.

Among them was Sidney Peel, one of the new volunteers from

Oxfordshire, who with telling insight recorded the motives of his fellow recruits in the Imperial Yeomanry:

> "Some came because they saw a chance of emigrating at government expense; some for love of sport and excitement; some because their domestic affairs were in a tangle from which enlistment offered a ready escape; some because they were tired of their present occupation; some because they wanted a job; some because they wanted a medal, and some because others came."

Peel remembered only one man who claimed to be motivated by patriotism and he was "generally regarded as peculiar"

"Trooper 8008"

The Hon. Sidney Peel, barrister and fellow of Trinity College, was the third son of Viscount Arthur Peel of Sandy, Bedfordshire, formerly Speaker of the House of Commons. He was perhaps an unusual man to find among the private soldiers, even in the Yeomanry, but his lack of previous military experience and his willingness to accept even so humble a position for the chance of the adventure, placed him in the ranks. There he recorded his experiences and those of many others like him in a book "Trooper 8008", published in 1901 immediately after sickness brought him home. He had no doubt about his own good fortune in being able to join the IY: "I, a mere civilian, went on active service without any of the bother of military education and peace training. I realise that we were much to be envied". Such attitudes contrasted strongly with the views of many regular soldiers and some at the War Office who felt that this lack of training and "the bother of military education" reduced the usefulness of Yeomanry on active service. There was a good case, however, that the unconventional background of Yeomanry, almost untouched by training which would knock them into the same mould as the rest of the regular army, made them particularly effective as irregular mounted infantry, the role for which commanders in the field like General Buller had originally requested them. He saw them as a match for the similarly untrained but skilful irregular forces of the Boer farmers. Whether they proved so is open to doubt, but they certainly made themselves useful as scouts and outriders on the open veldt of South Africa, skirmishing on the flanks of marching columns, and pursuing the fast moving Boers. They were also involved in a number of larger engagements, and in garrisons and patrols in the later, more static stages of the war of attrition waged against the Boer commandoes and their families by Roberts and Kitchener.

It was at Dreifontein, where the Oxfordshire men were involved in attacking and seizing a Boer position, that one of their officers was killed. Cecil Boyle had joined the QOOH in 1886, when like a number of its officers he gave a London address. He rose from Lieutenant to be Captain in "A" Troop before volunteering for South Africa. Other QOOH officers played a leading part in the administration of the Imperial Yeomanry during its first year in South Africa. Most notable among these was Lord Valentia who became Assistant Adjutant General, while the Duke of Marlborough also served on the general staff as a Captain. Both these two, together with Major Eustace Fiennes, SSM Goldie a former 3rd Hussar from Woodstock, and Sgt Barnsby were mentioned in dispatches.

Peel, from the ranks, sometimes expresses the frustrations of the private soldier in any war, who is at the receiving end of apparently pointless and contradictory orders and exhausting marches which seem to achieve nothing. "At the time we heard nothing whatsoever of the object of our operations. To us it was an incessant round of early reveilles, long marches, night guards and pickets, apparently without any aim or purpose whatever." When his superiors actually intended to keep their movements secret, he seems to have soon picked up the expertise of the hardened soldier in finding out what was going on: "Day after day reveille would be announced for 6 or 6.30 a.m. and regularly altered later in the evening to 3 or 4. Presumably the idea was that the Boers would be deceived by this, but as the real hour of starting was always to be ascertained by asking the headquarters cook what time breakfast was ordered, I have my doubts as to the efficiency of the plan."

He describes vividly the experiences of so many yeomen like him, adjusting to the hardships of active service, of long treks in difficult conditions and the joys of receiving letters and parcels from home—especially one which contained a thick green flannel shirt. "I shall not easily forget the shirt I got that mail . . . . I had worn my only other for more weeks than I care to mention, but it was not the weather to discard an old friend, so I wore the new one underneath it". This might have been one of many such comforts sent out to the troops from people in Oxfordshire, who collected funds and wrote letters to them. One of the most energetic was Sarah Angelina Acland, daughter of Sir Henry Acland, the former Regius Professor of Medicine at Oxford.

In spite of the hardship and the unpleasantness of some of the places he was forced to camp in—"a more foul and ill-smelling place than those outskirts of Kronstad I never wish to see"—the yeomen never lost their enthusiasm for the war or their confidence in the outcome. "As for defeat, it is unnecessary to say that none of the troops with whom I marched ever contemplated the possibility of it for one moment. . . .once it came to fighting we were perfectly certain to win." Unfortunately, one of the main events of their war was the failure to rescue a force of Imperial Yeomanry lured into a trap and forced to surrender to the Boers at Lindley.

The Lindley Affair

Peel, and many others in the army and War Office, blamed this on confusion and the slow response of the senior commanders. It seems that towards the end of May 1900 the Boers managed to trick Colonel Spragge, by means of a forged message purporting to come from his superiors, into taking his force of IY troops to Lindley, where he found himself cut off and surrounded. Lord Methuen, and General Colville commanding the 10th battalion (including the QOOH men of the 40th company), received his request for assistance but inexplicably took two days to cover the comparatively short distance in going to his rescue. They arrived too late to prevent the surrender of Spragge's force of 500 men and missed further opportunities as they chased after the retreating Boers. "Deep anger and mortification filled our hearts as we gazed upon the pitiful scene and realised that we had come too late."

Perhaps the men were still reflecting on this at a church parade on the open hillside overlooking Lindley which has been vividly recorded in a large painting of the scene, as well as in Peel's words. "The next day was a Sunday. We had a Church parade on the slope of the hill, and we had a sermon, of which I did not hear a word, and of course we sang 'Onward Christian Soldiers'. All the time, and indeed all the day, there was the occasional sound of rifle firing from the outposts; but it was not our turn to go out and we spent a quiet day. The service was conducted by a corporal of Yeomanry, who was, I believe, a curate by profession."

The Yeomanry in their element

There were happier moments, when the mounted yeomen came into their own, galloping fast across the hills with the excitement of the chase to spur them on. "We collected behind a bluff of the hills of our side of the valley, then trotted out, formed into a long line, and so went at a gallop across the valley. What a comfort to have a real good gallop again after so many miles of trekking and walking or the humblest of trots! From the hill behind us our guns thundered a

13. **Lord Chesham (centre) with officers of the QOOH at Blenheim after the annual inspection, 1901. Viscount Valentia and Winston Churchill wear medals and South Africa style hats, denoting their recent service in the Boer War. (Winston Churchill also appears to be wearing a different style of jacket and sporting a bow tie!). Also in this group are the 9th Duke of Marlborough (seated on ground) and "Jack" Churchill, Winston's brother (back row, left).** *(Oxfordshire Yeomanry Trust)*

most inspiriting music; the shrapnel screamed and whistled overhead. There is nothing more enlivening than hard riding to the tune of your own artillery. Down into the donga we went, and out again faster and faster. It was impossible to resist shouting as we rode."

After the eventual defeat of the Boer armies, the war deteriorated into one of blockhouses, barbed wire and concentration camps as remaining Boer resistance was ground down. Fewer QOOH men took part in the closing stages, most having served for a single year. Indeed Valentia returned in time to receive the Freedom of the City of Oxford at a ceremony in his honour on December 6th 1900 (he had been Oxford's MP since 1895). The Duke of Marlborough and other officers were back with the QOOH for the annual training camp at Blenheim in 1901, and in July both Valentia and Marlborough were among a prestigious gathering, mostly peers of the realm and including Field Marshal Lord Roberts, who gave a dinner in honour of Lord Chesham at Londonderry House.

The aftermath of war: Haldane's army reforms

The Boer War ended in victory for Britain, but at considerable cost in lives and money. In spite of huge public celebration of events like the relief of Mafeking and of the final victory, in private there was a good deal of soul-searching about the efficiency of the army, and especially the effectiveness of volunteers. The pressure for national conscription on the continental model increased in some quarters and was only resisted by Lord Haldane's proposals for reform of the army, finally introduced in 1908 after much argument. Earlier attempts to reform the Yeomanry by retaining the title "Imperial Yeomanry" (even wearing South African style slouch hats for a time), and making them into mounted infantry armed with rifles and bayonets instead of the traditional cavalry sword, met with such vociferous opposition that the measure was abandoned after being in force for only a few years.

When they were finally adopted, Haldane's reforms reorganised all the former auxiliary forces, militia, volunteers and yeomanry, into a new Territorial Force with a mainly home defence role in any part of the UK, but also able to volunteer for overseas service. Enlistment was to be for three years, plus a further year on the reserve, and there would be 15 days annual training in order to overcome the previous deficiencies of yeomen who only trained for half that time. There would be six months full-time training if hostilities broke out.

14. **General Baden-Powell, hero of the seige of Mafeking, inspects the QOOH at their annual camp, 1902.** (*Oxfordshire Yeomanry Trust*)

Although there had been an increase in recruitment to the yeomanry and other volunteers in the wave of patriotism immediately after the Boer War, the new Territorial Force was not so popular and numbers remained below the expected targets. Even the longer training and the emphasis on practical skills such as marksmanship did less to improve efficiency than had been hoped, which probably explains the ignominious result of a shooting contest reported in the Banbury Guardian in 1911, when a Britannia Works team beat the Banbury Squadron of the Queen's Own Oxfordshire Hussars 706 to 654.

There were still great occasions to be enjoyed, however. The QOOH sent a detachment to represent them at the Coronation of

Edward VII (still their Colonel in Chief). This was the year in which the annual inspection was carried out by that great hero of Mafeking, Major-General Baden Powell. He observed a training scheme based on the somewhat unlikely scenario that "an invading force having landed in the South of England and taken possession of London, is advancing northwards in three columns the headquarters of which has reached Witney, Oxford and Thame", while the remaining defenders of England were conveniently bivouacked in Blenheim Park. In 1907 Lord Haldane himself, accompanied by the Chancellor of the Exchequer Mr Lloyd George, visited the annual camp at Thame. Other important visitors included the King of Portugal, perhaps through a connection with Major James Mason of Eynsham Hall who had come to the notice of the king during

15. **Influence in high places. In 1907 when Haldane's proposals for the reform of the Auxilliary Forces were being considered by the government, Lord Haldane himself, Lloyd George (Chancellor of the Exchequer) and their ladies were entertained by the QOOH at their camp at Thame. Also in the picture are (front row from left) Winston Churchill, F E Smith (2nd Lord Birkenhead), the Duke of Marlborough, Sir R. Hermon-Hodge M.P. and Maj. Eustace Fiennes.** *(Courtesy of the National Army Museum)*

his remarkable career as an engineer and owner of rich copper mines in Portugal. There were glittering annual dinners in London for the officers, and for the men a special concert was arranged one year in Oxford Town Hall, when leading actresses and performers from current London shows were hired to entertain the soldiers. Regimental pride and esprit de corps were high. They were about to be put to the test as never before.

V Flanders, 1914–1918

August 1914, Mobilisation

From very early in the morning on Wednesday 5th August 1914 men of the QOOH were arriving in the pouring rain at their squadron HQs in Banbury, Oxford and Henley to report for wartime duty. They came in answer to a telegram received the previous night bearing the single word "Mobilize", and followed a routine set up as long ago as 1909 for the use of the Territorial Force in time of war.

Many things were to change from that day on, and for the men of the Queen's Own Oxfordshire Hussars the contrast between the history of their regiment over the first hundred years of its existence and its forthcoming experience in the second, could hardly have been greater. From being a largely social institution with little more than a veneer of military expertise for much of the nineteenth century, it was to become part of the professional army in a very real sense, seeing action in both world wars, suffering many casualties and sharing the hardships of war to the full. An army regiment always has two sides to its life—as a social organisation where class distinctions, glittering full-dress uniforms, grand ceremonial parades, sporting activities, regimental traditions and a sense of belonging are all important, and its other existence as an efficient fighting unit in wartime. The two are not really as separate as they seem, each contributing to the other. The preparation for the

16. **The Woodstock Troop shortly before the First World War.** (*Oxfordshire Yeomanry Trust*)

second role had begun for the QOOH with the reoganisation of the Territorial Force by Lord Haldane, the extended training camps and the greater emphasis on military efficiency. The QOOH no longer operated on its own but had for some years been part of the 2nd South Midland Mounted Brigade, and had within its ranks several ex-regular officers and NCOs and others who had seen service in South Africa. The career of its Commanding Officer spanned the period of change. Arthur Dugdale who lived at Sezincote near Moreton in Marsh, had joined 'D' Troop of the QOOH as 2nd Lieutenant in May 1892, serving under five other COs before himself taking command in 1914. He was to keep that role throughout the war.

The Churchill influence again—embarkation

A further period of intensive training followed mobilization, with much laborious movement of men and horses from Reading to various camps in eastern England before the whole brigade, men, horses and equipment, moved to Churn on August 30th. (This operation took no less than twelve trains leaving at intervals between 8a.m. and noon for the hot eight-hour journey via Cambridge, Bletchley and Oxford). Training continued for a fortnight or more and then, in the early hours of September 19th, an amazing telegram was received from no less a person than the First Lord of the Admiralty ordering the embarkation of the QOOH for France.

The First Lord was of course Winston Churchill, energetic member of the QOOH since 1902, though in recent years his political career had left him little opportunity to indulge his enthusiasm for the yeomanry, and for this local regiment in particular, in which his brother Jack was serving as a major and second in command. Now in his boyish way he seized the opportunity for his regiment—inexperienced Territorials though they were—to have the first taste of a great adventure by arranging for them to go to France in support of his Naval Division. They thus enjoyed the honour of being the first Territorial regiment to embark for active service in the war. Once there, while some of its senior officers were involved with the Naval Division supporting the Belgian Army at Antwerp, the rest of the regiment was left very much to its own devices and busied itself in operations behind the lines, at one point lending support to a French unit and experiencing their first exchange of gunfire with the enemy.

The role of cavalry in war was to exploit its potential for moving quickly, in flanking movements or in pursuing the enemy when other forces achieved a breakthrough. It had been at its best in the open country of South Africa. It proved to be a very different story for the QOOH in the mud of Flanders, when the war on this front ground to a virtual halt amid the trenches, and there was little scope for the cavalry to use its skills. There were consequently several periods of being held in reserve behind the lines, waiting "to go through the gap", which was always about to be created by the infantry and artillery and through which the cavalry hoped to pursue the retreating Germans. In the early stages of the war these periods were put to good effect by training the inexperienced members of the regiment and preparing them for eventual action. Swords had been issued to the men on mobilization, but it was to be a war of rifles, bayonets and machine guns and much practice was needed. There was also plenty of practice in digging trenches and there were endless guards to mount and fatigue parties to supply.

There were also opportunities for recreation of one sort or another. Some officers had even brought private cars over on the troopship and indulged in joy-rides around the area until one party casually drove up the Menin Road and nearly got caught up in the front line! There was hunting with beagles brought over by another regiment and polo which the QOOH pioneered in France. There was football for the men and at Christmas presents provided from a fund organised by the CO's wife as well as the famous tin box from Princess Mary to all the troops, containing tobacco, cigarettes and a card.

Messines, October 1914

The first painful experience of the realities of war had come at the end of October 1914, with the men dismounted and fighting in the countryside around Messines. Apart from the shells and bullets of the German guns, Lieutenant Keith-Falconer's account describes the discomforts of hunger, thirst, cold and weariness. "A little bully beef and biscuit and a mug of tea were all that the great majority, including officers, had had since leaving St Omer two days before .... By the time we got to billets that night everybody had been over

17. **Officers of the QOOH relaxing at camp in the summer of 1914.** (*Oxfordshire Yeomanry Trust*)

sixty hours without sleep." Captain Molloy and Private Sheasby were killed and several others wounded in this action, but Sir John French congratulated the troops involved who had held the Wytschaete-Messines Ridge for almost 48 hours against much larger forces "The greatest threat of disaster was staved off by the devoted bravery and endurance of the cavalry corps." Messines was deservedly the first battle honour of the war earned by the QOOH.

From Ypres to Arras

For the next three years the QOOH shared the general experience of all troops fighting in Flanders: extreme discomfort, mental anguish, fear and bravery, excitement and boredom, the comradeship which comes from shared dangers, and confidence in the final outcome. During the second battle of Ypres they witnessed a gas attack "a thick greenish haze coming over the ridge towards our trenches", and endured a continuous artillery bombardment for more than two hours. By the end of May 1915 they had lost 20 men with a further 45 wounded.

Keith-Falconer recounts that one of the worst duties was taking supplies up to the trenches at night. "The enormous weights to be carried—heavy boxes of ammunition, biscuits and bully beef besides of course each man's own rifle and ammunition, and many other things such as parcels, rum jars, extra entrenching tools, etc.—the darkness and difficulty of keeping in touch, the rough ground pitted with shell holes, the extreme probability of being shelled or sniped at or both." Some claimed they would rather be in the trenches than have to endure this excessive fatigue of support duty. Even behind the front line men and horses were affected by the weather. "It rained almost incessantly for several days. The scene of desolation presented by the horse lines was unimaginable: a stretch of ground about a mile long by 400 yards broad, every available inch occupied by horses . . . . and the whole surface churned into a state of mud and slush where men reeled and staggered as they walked, sinking over their ankles at every step . . . . The men and the horses were in a pitiable condition. Nearly all the men had got tents or shelters of some kind, but few had mackintoshes or any change of clothing, and their 'British Warms' soon got sodden through. Also it was practically impossible for them to keep their blankets dry . . . . As for the horses they looked and were miserable." In spite of such conditions, or perhaps because of them, it was still important to keep up morale, and each year in July the Brigadier gave a dinner for Old Etonians when several officers from the QOOH attended if available. For the men there were less refined amusements: in July 1917 during a rest period, sports were organised which included a race in which competitors had to run 50 yards then drink a mess-tin of beer, undress and run back and light a cigarette before returning to put their clothes on again. Such events afforded a brief respite from reality: three days later 20 men were killed or captured on a night working party.

The battle of Arras started with excitement for the cavalry in anticipation of the hoped for breakthrough in the German lines. Once more, however, it ended in disappointment as no break occurred, and after three days poised for action, the men and horses returned to their billets, tired and frustrated. There was considerably more action, if on a smaller scale, for the QOOH at Gillemont Farm in May 1917. Major Valentine Fleming was in command of a squadron holding this rather exposed outpost where there was not the usual continuous line of trenches. On the fourth night there his position was fiercely attacked by perhaps two companies of German infantry following a bombardment with trench mortars and heavy guns. Major Fleming, another officer and four privates were killed, but the attack was repulsed by the determined efforts of the remaining men and two prisoners were captured. Major Villiers, who came up in support and showed great personal courage was later awarded the D.S.O. There were a further sixteen casualties and many wounded at this post in the next few days.

1918—German offensive and the attack on Rifle Wood

March 1918 brought what proved to be the last German offensive and in its early stages there was near chaos as the outnumbered British fell back. Men of the QOOH were again dismounted and often split up on different assignments. It was a time of continual movement from one position to another, urgently digging-in and then being moved elsewhere. On one occasion machine guns, a Hotchkiss rifle and its ammunition had to be abandoned because the transport went to the wrong place in the chaos. Orders were constantly being given and then cancelled. For the men there was little time to rest or eat and the overall situation was very unclear. As the desperate need for more men to hold the line increased, those left in the rear to look after the horses of the QOOH were brought into action so that at one time the horses of each troop had to be roped together in a long line "like beasts going to market" with a man at each end to control them.

The German advance was eventually held back by numerous actions along the front, like that at Rifle Wood near Amiens in which the QOOH played a leading part in cooperation with the 3rd Hussars and other regiments. A determined attack by three waves of troops on the German position in this wood succeeded in driving the enemy out and then holding the position in spite of a heavy bombardment by artillery and the difficulty of digging trenches among the trees while under fire. Twelve men of the QOOH were killed and a further 50 were wounded of whom two later died, but it was considered a very successful operation which brought warm commendation from the Divisional Commander: "The attack was carried through with dash and gallantry and reflected the greatest credit on all ranks, especially when it is considered that the Division had been in action since the 21st March, during which time they had suffered very heavy losses and had not had time for rest and re-fitting".

The final advance

As summer approached the pressure eased a little and the Allies began to gather their own strength for the final advance and the long-awaited breakthrough in the autumn of 1918. On one day that summer Mr Winston Churchill, then Minister for Munitions, "dropped down from the sky in an aeroplane" when on his way to a meeting at GHQ, and lunched with the officers of the Oxfordshire Hussars.

There was still plenty of fighting to do, and casualties to be suffered, but when the advance started in August it saw the cavalry at last restored to their proper role, and able to move quickly across open country. There was a sudden demand for cavalry to support the infantry. "The moment the war became one of movement, every unit was screaming for mounted troops to help them, their previous opinion of their uselessness having suddenly changed." The advance continued throughout September and October as the Germans began to fall back on all fronts, and on November 11th the final armistice was signed.

The regiment which returned to England early in 1919 was very different from the one which had gone so unexpectedly to Flanders in the first months of the war. Whereas it had formerly been derided by regulars as the "agricultural cavalry" and its QOOH title used to create the nickname "Queer Objects On Horseback", now, hardened by four years of gruelling active service and raised to a degree of military efficiency previously unheard of in a yeomanry regiment, it was almost indistinguishable from regular cavalry. In addition to official commendations, perhaps the clearest recognition of their new status came from the officers of no less a regiment than the 3rd, The King's Own Hussars, alongside whom the QOOH had fought for much of the war, who hoped that "the Officers of the Queen's Own Oxfordshire Hussars, past present and future, will always consider themselves Permanent Honorary Members of their Mess".

VI Territorial Artillery Between the Wars

Territorial politics

As the men of the Queen's Own Oxfordshire Hussars were finally returning from Belgium to Catterick Camp and from there to resume their civilian lives in Oxfordshire, the role of the Territorial Army again became a political issue. There would have to be a huge reduction in the size of the armed forces and questions arose as to the effectiveness of a part-time force which could not be used outside the UK except in a national emergency and after formal parliamentary approval had been given. The old rivalries between professionals and amateurs were revived. The government wished to have complete control of all army units, without restriction on their organisation or deployment, while the Territorials were anxious that they should retain their integrity as individual units, even in an emergency, and not simply be drafted into other regiments.

Fortunately for the Territorial Associations the new Secretary of State for War was Winston Churchill, that old enthusiast for the amateur tradition. Although the Territorials did not like every detail of the new arrangements, Churchill pledged to preserve the integrity of the units and retained the restriction on their use abroad. (It was also typical of Churchill that he tried to revive their role in support of the civil powers in case of strikes and disturbances, etc., but this was resisted by government to the relief of most Territorials).

A new role for the Yeomanry

In some respects it was the Yeomanry who came off worst in the new arrangements. It had become obvious that cavalry would have only a limited use in modern warfare and the number of mounted Yeomanry regiments was drastically reduced. Of the fifty-six existing when the war ended, only fourteen survived the 1922 reorganisation. Sixteen were disbanded altogether and the remainder were converted to other roles. Among the latter were the QOOH, who became an artillery unit from that date. There was considerable regret and some resentment at this decision, being based as it seemed on an unfair assessment of the seniority of the regiment. The QOOH traced its history back to the Wootton Troop raised in 1798, and later incorporated into the First Regiment of Oxfordshire Yeomanry Cavalry, but because this regiment was on the list of those to be disbanded in 1827, although the king allowed it to continue to serve voluntarily without pay or allowances, its seniority was only reckoned by the War Office from the date of its return to full status in 1830. "Older" regiments of Yeomanry retained their cavalry status but the Oxfordshires became gunners.

For some this must have felt like the end of the Oxfordshire Yeomanry. Captain Keith-Falconer certainly expressed such a sentiment in his account of the regiment in the First World War, written at about this period. He wrote that the regiment was "finally disbanded and ceased to exist as a military unit" in March 1922. Referring to the new batteries of field artillery which replaced it he said that they were "recruited mainly from the towns and must of necessity develop a new spirit and a new tradition, very different from that of the old Yeomanry cavalry". He was alluding to the feeling of many

18. **Horse-drawn artillery between the wars. One of the QOOH Batteries training as part of the 100th Worcestershire and Oxfordshire Yeomanry Brigade, Royal Artillery (TA).** (*Oxfordshire Yeomanry Trust*)

country dwellers that they came from a different background and had a different set of values from townspeople (a feeling still being expressed by some in the late 1990s). The traditional social hierarchy of rural society was still based on the landed gentry and the farming interest, whereas in the towns middle-class manufacturers and tradespeople tended to be more prominent. Hunting seemed to represent the essence of the rural scene, and mounted cavalry with swords were the military expression of this. Before 1914 there had been many comments about the suitability of officers based largely on their hunting background. In the new artillery horsemanship in the old sense would be of less value than technical skills in gunnery, and these were not the sort of skills generally valued in traditional rural society.

"Yeomen first, Gunners second"

Whatever the truth of these sentiments, many features of the old Yeomanry regiment were carried over into the new force. A number of the officers did transfer including Keith-Falconer himself, who was later promoted to Major and became one of the Battery Commanders. Horses were retained to haul the guns until motor vehicles became available and officers and headquarters staff continued to ride horses for some years. The title "Queen's Own Oxfordshire Hussars" together with the badge bearing Queen Adelaide's cypher were still used although the men now formed the 399th and 400th Batteries of the 100th Worcestershire and Oxfordshire Yeomanry Brigade, Royal Artillery (TA), and stories still circulated about officers who were solicitors or accountants in private life giving casual and decidedly un-military replies to fiery inspecting officers during the annual camps. Winston Churchill remained Honorary Colonel and Queen Mary continued as Colonel in Chief, having accepted that role in 1919. A commanding officer who served much later than this confirmed that the men of the regiment continued to think of themselves as "Yeomen first and gunners second".

Annual Training Camps

The regime of training for Territorials consisted of one evening a week at a local centre and one annual camp of fifteen days under canvas. These were usually held outside the county and often at one of the army's gunnery ranges on Salisbury Plain, where practical training was given in the handling and firing of 18-pound field guns and 4.5 Howitzers (rather ancient ones with wooden wheels according to one account). Churchill once attended a camp but having seen the conditions took up his quarters in the local inn. There he learned the rudiments of gunnery over dinner with the Adjutant and the Instructor, and acquitted himself surprisingly well commanding a battery on the ranges next day. Every two years a gunnery competition was staged on the ranges among TA artillery batteries for the King's Cup. The QOOH twice reached the final and on one occasion won it outright.

Important as the camps were for serious training they were also something of a holiday—and for many of the men the only holiday they got in the year. Away from work, and home, under canvas in the company of other men of similar age and background, there was plenty of fun to be had as well as work to be done, and the men undoubtedly looked forward to this annual experience. Opportunities for recreation were increased in alternate years when the so-called "holiday camps" were held instead of "firing camps". In the harsh conditions of the depression between the wars, governments were forced to economise on the cost of the TA and so in alternate years the camps did not involve firing with live ammunition. Consequently parades tended to finish at about midday and those not on duty went into the nearest town to enjoy themselves. These were also harsh times for many individual working men and for the growing numbers of unemployed. Many were attracted into the TA, especially in the industrial areas, by the prospect of the annual camp and its army rates of pay with all found. Even apart from the camps there was a social programme centered around the drill hall in the home town, consisting of football, cricket, boxing, various excursions for the families of yeomen and regular dances.

Queen Mary in Oxford

One of the last peacetime duties carried out by the QOOH provided a real link with its past. In June 1937 Queen Mary paid a visit to Oxford in order to lay the foundation stone of the New Bodleian Library on the corner of Broad Street and Parks Road. She was still Colonel in Chief of the regiment and they proudly provided the royal guard of honour, just as they had done for Queen Adelaide a

19. Queen Mary going in procession to the Sheldonian Theatre to receive an Honorary Degree from Oxford University, 1921. In the background is the mounted escort provided by the QOOH, of which she was Colonel in Chief. (*Oxfordshire Yeomanry Trust*)

hundred years before. At least one officer present wore a borrowed pair of the famous pink-heeled boots designed in the days of Lord Valentia and the Prince of Wales.

The 63rd Anti-Tank Regiment

Another change of role came in 1938. War with Germany was looming again in spite of all efforts to avert it, and there was a demand for increases in the armed forces although the idea of reintroducing conscription was still unacceptable to many politicians. Prime Minister Chamberlain decided to double the establishment of the Territorial Army and this resulted in the QOOH becoming a separate regiment. Their new role was as an anti-tank unit and they were redesignated the 63rd (Oxfordshire Yeomanry) Anti-Tank Regiment RA (TA). The "TA" addition was soon to disappear because as war became inevitable in the following year conscription was introduced, restrictions on conditions of service were cancelled "for the duration" and all units became part of the national army.

VII THE SECOND WORLD WAR

Second Line duty

If the men of the QOOH were as keen to get into action when the Second World War started as they had been in 1914, they were to be disappointed. Once again the regiment's lack of historical seniority went against it, and it was given only a Second Line role. When all the Territorial Army regiments were "doubled up" as a preparation for war, the Worcestershire and Oxfordshire Yeomanry Batteries were separated as well as being increased in size. The Worcestershires became the 53rd Anti-Tank Regiment and were posted to France, while the Oxfordshire batteries formed the 63rd Anti-Tank Regiment and moved only as far as Warwickshire in the first few months of war. Like other Second Line units its role was to train men to feed into First Line regiments but it also had a vital part in defence against invasion. This role would eventually be taken over by the newly formed Local Defence Volunteers (Home Guard), but they would need time to be trained and organised. Meanwhile regiments like the Oxfordshire Yeomanry were moved into strategic positions on the south coast immediately after the evacuation from Dunkirk, when it was expected that Hitler would press home his advantage by attempting an invasion of a weakened England. The threat did not materialize, however, and there were soon new postings for the regiment even further away from action in the front line.

Ireland

Even when the immediate threat of an invasion of England receded it was still felt necessary to station a defensive force in Northern Ireland, in case the enemy should attempt to land troops in neutral Eire and attack England through her "back door". The Oxfordshire Yeomanry were among those posted there as part of the 61st Division in June 1940, most of them remaining until February 1943. While to some men (and undoubtedly to their families at home) this must have been a relief, to others it no doubt brought a sense of frustration and perhaps guilt at not sharing in the hardships and risks of war born by front line troops. Perhaps it was hard to accept that their own role was also vital to success. The only course, however, was to make the best of their situation and the Yeomanry set about doing this.

Their first priority was to raise their own level of efficiency and preparedness to take on whatever role the war thrust upon them. The 61st was perhaps one of the few fully equipped divisions at this time because so much had been left behind in the scramble to get the troops safely out of Dunkirk. Time was now spent in making the men, including many new recruits as well a those with pre-war territorial experience, proficient to wartime standards in all aspects of gunnery. There were also more mundane skills like vehicle maintenance to be mastered. Training courses were provided for all ranks and there were frequent exercises, sometimes involving the whole division. One officer recalled that when access to the live ranges was not possible, practice was improvised by firing tracers at a mock-up tank from a rifle fixed on top of the gun's barrel. As well as the routines of army life there were sports and entertainments to keep up

20. **A hungry crowd at the cookhouse. Men of the QOOH in Northern Ireland during the early years of World War 2.** (*Oxfordshire Yeomanry Trust*)

fitness and morale both in and out of camp. Some officers even managed to enjoy grouse shooting and fishing, and periodic leave was possible for all ranks. There were postings in and out for individuals from time to time, and in September 1941 the whole of 251 Battery (one of the "Banbury" batteries) was sent back to England to join a force being hastily put together for service in the Far East, in answer to the growing Japanese threat. A very different experience awaited them.

Singapore and the Burma Railway.

Since 1937 Japan had been fighting a quite separate war against China and now saw the chance to strengthen her economic and military position by seizing former British, French and Dutch possessions in South East Asia. Having destroyed much of the American Pacific Fleet at Pearl Harbour in December 1941 before it could be used to oppose her, the Japanese armies moved into Malaya taking advantage of British preoccupation elsewhere. Their planes sank two British battleships and together with the land forces threatened the main British base at Singapore. It was against this desperate background that Churchill belatedly diverted troops intended for the Middle East to Singapore instead. Among them were the men of 251 Battery, now part of a different anti-tank regiment but proudly retaining their Oxfordshire Yeomanry title.

Singapore fell on February 15th and these men found themselves among the 60,000 who surrendered and were destined to spend the

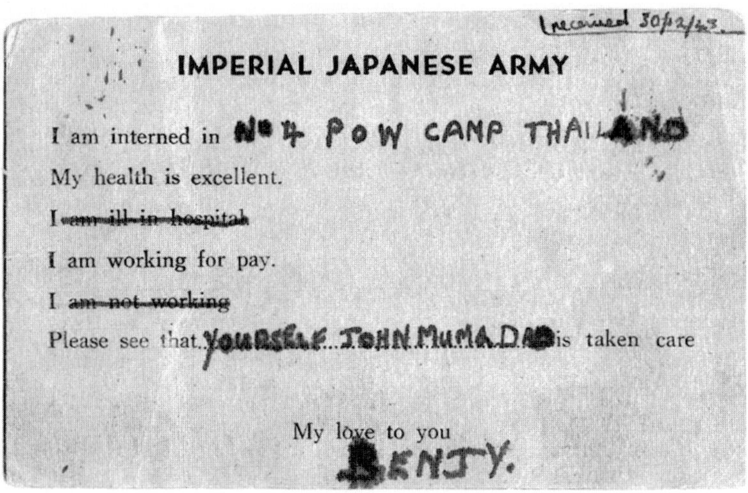

received 30/12/43.

IMPERIAL JAPANESE ARMY

I am interned in No 4 POW CAMP THAILAND

My health is excellent.

I ~~am ill in hospital~~

I am working for pay.

I ~~am not working~~

Please see that YOURSELF...JOHN.MUM.DAD is taken care

My love to you

BENJY.

21. **Approved style of postcard sent to his wife by BSM Ben Pritchett, a POW working on the Burma Railway, December 1943.** (*Oxfordshire Yeomanry Trust*)

rest of the war as prisoners. For three and a half years they suffered the most appalling cruelty and degradation while being forced to work on the construction of the notorious Burma-Siam railway. Over 16,000 died before the war ended. Only those who went through this experience can describe its horrors, and most choose not to do so.

The Yeomanry in Europe: the Churchill connection again

The rest of the regiment remained in Northern Ireland until February 1943 when they moved back to England. There followed a further period of training in large-scale exercises sometimes involving American and other troops stationed in England in readiness for the Normandy landings. For a time in 1944 men of the Yeomanry were used to set up and staff a large camp for 3,000 US army personnel. When "D Day" came on June 6th the only men left behind in the camp were the Oxfordshire Yeomanry. It was rumoured that they were to be dispersed to make up the numbers in any frontline units that needed more men.

This situation was too much for the Commanding Officer. John Thomson had first joined the regiment as long ago as 1927 and had worked his way up from 2nd Lieutenant to become its CO in April 1943. He was thus well acquainted with its traditions and connections. The most important of these connections was Winston Churchill himself, Prime Minister but also Honorary Colonel of the regiment. Another was the second Earl of Birkenhead, who had been an officer in 252 Battery and was a godson of the great man (his father F E Smith having served with Churchill in the QOOH and in government). With Birkenhead's help an approach was made to the Prime Minister asking him to use his position to get the regiment into action, just as he had done on the famous occasion in 1914.

Not surprisingly there were immediate results, and on the Prime Minister's personal orders the 63rd (Oxfordshire Yeomanry) Anti-Tank Regiment was moved to Berkshire, then in October to Newhaven, and a few days later embarked for Dieppe. On arrival they faced two problems: one was that all their guns and heavy equipment had been left behind in England and for about a month they were unable to play any very active part. The second problem was that the circumstances of their arrival in France through the personal intervention of Churchill made them unwelcome in some quarters, apart from the fact that nobody actually needed an extra anti-tank regiment. Now part of the Second Army, they were disappointed to be told that they must amalgamate with another regiment, the 91st (Argyle and Sutherland Highlanders) Anti-Tank Regiment. This meant that half the men in each regiment were dispersed elsewhere, to their great dismay. It was further decreed that neither of the two COs could command the new unit, and so John Thomson was also posted away.

For the next few months the Oxfordshire Yeomanry batteries and their new Scottish colleagues followed the advancing armies across Northern France, Belgium and into Germany as the enemy were driven back. There was hard fighting at the front and a number of engagements even in the second line which brought casualties to the Oxfordshires. At last they were able to put into practice the training they had received for so many years.

The relief of Belsen

No amount of training, however, could have prepared them for the terrible experience which now faced them. In April 1945 the regiment was given the task of taking over from the Germans the concentration camp at Belsen. Colonel Taylor, Major Barnett and the men of 249 Battery were the first British troops to enter this camp and to observe its horrors. It was thought by the German SS Commander to contain about 60,000 people. Huge piles of unburied corpses lay on all sides, and those still alive were little more than terrorised walking skeletons suffering from starvation and disease. The Oxfordshire men spent over a week assuming control of the camp and attempting to alleviate the suffering of the internees until medical and other staff could be sent up to take over. The experience was certainly traumatic. The true nature of Belsen was publicised to the world by reporters and photographers sent there for that purpose by the army command.

The final months of war

The last posting of the war was to the former German naval base of Kiel on the Baltic coast. Here the Yeomanry enjoyed a less stressful time guarding the base and its naval stores and rounding up "displaced persons".

After the eventual defeat of Germany and the surrender of Japan in 1945, those who had been fortunate enough to survive were finally demobbed, the men imprisoned in Thailand were freed, and all returned to enjoy the fruits of peace.

VIII The Yeomanry Since the War

The return of peace

The return of peace after the Second World War once again raised the question of the role of the Territorial Army in peacetime, just as it had at the end of the Boer War and the First World War. The changing nature of warfare in the second half of the twentieth century and the impoverished state of Britain as a result of its huge expenditure during the war would be vital considerations, but for the time being the difficult issues were shelved until proper long-term plans could be drawn up for the armed forces as a whole. In 1947 the old TA units were re-formed much as they had been before the war and this resulted in the QOOH becoming 387 Field Regiment, Royal Artillery. A number of men who had served throughout the war rejoined, including even Major Turrill who had suffered so much in Japanese prison camps that he might have been expected to turn his back on soldiering in future. Another link with the past was that their commanding officer was once again Lt. Colonel John Thomson, reunited after his forced separation from the regiment in the last years of the war, and intent on keeping alive the Yeomanry tradition. He commanded the QOOH until 1950 but continued his connection with them for many years after that as

Honorary Colonel, first of the QOOH Battery then of 299 Regiment RA of which they were part until their disbandment in 1967. Through successive reorganisations of the TA he was Honorary Colonel of the Oxfordshire Territorials and of the Oxfordshire Cadre up to 1975 and President of the Oxfordshire Yeomanry Old Comrades Association until his death in 1998. He managed to devote so much time and energy to supporting the Yeomanry in spite of an exceptionally busy private life as Chairman of Barclays Bank from 1962 to 1973. He also served as Lord Lieutenant of Oxfordshire for a time, thus reviving the historic link between that office and the Yeomanry.

Reforms and amalgamations

By 1950 the government was ready to publish its plans for the future organisation of the army, based on the new post-war situation and the expected role of Britain in Europe and the world. An important factor was also the huge cost faced by Britain in rebuilding all aspects of national life after the war, and it was inevitable that there would be a great reduction in the Territorials. Many units disappeared and others were amalgamated. For the QOOH it meant amalgamation with their neighbours the Royal Bucks Yeomanry to form 299 Field Regiment and six years later they were joined by the Berkshire Yeomanry. Separate batteries of this regiment were spread over the three counties and Oxfordshire's 'Q' Battery continued to carry the QOOH title, with troops based at Banbury and Oxford.

Churchill's funeral

Sir Winston Churchill, who had been a member of the QOOH since 1902, had first become Honorary colonel in 1927 and he retained that post until his death in 1965. Even then he brought one last honour to the regiment with which he had had such a lasting personal connection. One of the remarkable features in the story of this remarkable man was that he planned his own funeral, leaving detailed instructions for the conduct of proceedings on the day, which would of course be a great and solemn state occasion. A sealed copy of these instructions had for some time been held in a safe at the TA Centre in Oxford because they contained important orders for the Queen's Own Oxfordshire Hussars. It was to them that the supreme honour was given of marching ahead of the coffin at the state funeral procession through the streets of London.

For a TA regiment composed of part-time volunteers this was not only an honour but a considerable challenge. A detachment of thirty men was formed under the command of Major Tim May, the Battery Commander, and a regular drill Sergeant from a Guards regiment was quickly drafted in to train them in the ten days before the funeral. For those taking part, almost more moving than the funeral itself was the eerie rehearsal held a few days before in the early hours of the morning through the darkened streets of London, with no sound except the measured tread of the slow march and the beat of a single drum.

The tension and solemnity of the day itself was lightened by an incident absolutely in keeping with the independent tradition of the Oxfordshire Yeomanry. The QOOH role at the funeral was certainly unusual and one which no doubt aroused some jealousy and resentment among members of more prestigious Guards regiments who felt the honour of preceding the coffin should by tradition have been theirs. As the procession was forming up leaving space for the catafalque bearing the coffin, a very senior Brigade Major of the Guards triumphantly stormed up to Major May and indignantly pointed out that his men were arranged incorrectly according to state funeral protocol. He was completely deflated by May's reply that "In the Oxfordshire Yeomanry we always do state funerals this way", and retired in silence! It was a remark worthy of many of his noble predecessors and one which would certainly have pleased

Churchill himself, and if the QOOH are ever asked to take such a role again this particular arrangement of the men will no doubt have become a regimental tradition.

Further reoganisation: from gunners to signallers

The reforms of the 1950s were only the beginning of a major change in Britain's defence policy which in future was to be based on the concept of a nuclear deterrent. At the same time huge reductions were made in overseas commitments including the withdrawal from former British bases east of Suez. At home the army was further reorganised and the role of the Territorials severely reduced on grounds of cost as much as strategic necessity. Among the regiments to go was the QOOH which was officially disbanded in 1967.

This was in fact the third time in its long history that the regiment had been disbanded. It had happened previously in 1828 and again in 1922 but in the present case as in the past, the regiment refused to die. Within a very short time it was revived in an entirely new role when in 1971 the commanding officer of a Royal Signals squadron based in Banbury applied for permission to assume the title "5 (QOOH) Squadron", in order to perpetuate the Yeomanry tradition locally. This was a Special Communications unit supplying the needs of the Foriegn Office for secure communication systems in many parts of the world. A further Defence Review in 1974 reduced this role to home defence only but left the QOOH as the Banbury squadron of the 39th (Skinners) Signal Regiment whose headquarters are in Bristol.

The Yeomanry today

The successors to the original Oxfordshire Yeomanry now operate as a link in the National Communications Radio System with facilities for secure data transmission (DUST—Dial Up Secure Telegraph) or voice communication and a vehicle-based Combat Net Radio if required. Their official role is home defence but several men have seized the opportunity to serve abroad when called on to make up shortages in the regular army. Thus two men of the QOOH were sent to the Falklands and seven more have served more recently under the United Nations flag in Bosnia.

The modern Yeoman is a far cry from his nineteenth century predecessor. The modern "Yeoman" may indeed be a woman, as the army has accepted women in nearly all roles since the early 1970s. Major Heather Collins became the first woman to command the QOOH squadron in September 1992. All ranks are now trained specialists with technical skills beyond the comprehension of

23. **Members of 5 Squadron (QOOH), a Royal Signals TA unit, who are the present day successors to the Oxfordshire Yeomanry.** *(5 Squadron)*

volunteer cavalry officers in their "mantua purple" uniforms of a hundred years ago, when family connections, appearance and swagger were more important than military knowledge. Yet some things still link today's Yeomen to the past. One of 5 Squadron's duties, like other TA Signals units, is to provide military aid to the civilian administration when called upon to do so in any civil emergency or disaster, by establishing secure communication facilities via Instant Control Points. Though a different, and more acceptable, form of support, this is the exact counterpart of the old yeomanry tradition of helping the civil power to quell riots and tumults in the early nineteenth century.

Tradition is an essential element in regimental pride which inspires modern soldiers to give their best just as it always has done, and 5 Squadron has demonstrated its keenness to keep alive not simply the memory but the reality of the yeomanry spirit in recent years. Each new recruit receives instruction in the history of the QOOH and their uniform and badge are still in use. For many years veterans who served with the QOOH in France and Flanders held reunions at Middleton Stoney, the home of Major the Hon. Arthur Villiers, and there are still a number alive who fought in the Second World War. The famous banner listing battle honours from the time of the Boer War still hangs in the TA Centre in Banbury, and these veterans joined men and women of the present squadron to march proudly through the town with "bands playing, drums beating, colours flying and bayonets fixed" to receive the Freedom of Banbury from the Mayor in the presence of the Lord Lieutenant, in recognition of the Oxfordshire Yeomanry's bicentenary in 1998.

23. Ancient and Modern. Major Paul Wenlock, Commanding Officer of 5(QOOH)Sqn. in Royal Signals uniform, with L-Corporal Colin Bowdery wearing the traditional uniform of the Oxfordshire Yeomanry. Behind them is one of the two public houses in the county which are named after the regiment. (*Oxford Times*)

INDEX

(Figures in bold refer to illustrations)

The Author

David Eddershaw has lived in Chipping Norton, Oxfordshire, for more than thirty years. An historian and teacher by profession, he was for many years Education Officer and Assistant Director of the Oxfordshire County Museum Service. Now retired, he indulges his passion for telling people about the social history of Oxfordshire through lecturing and writing.

Other publications include "Using Local History Sources—a teacher's guide" (with James Griffin), Hodder 1996, and "The Civil War in Oxfordshire", Sutton 1995.